MW00844295

FEEDBACK AMPLIFIERS

Feedback Amplifiers

Theory and Design

by

Gaetano Palumbo
University of Catania

and

Salvatore Pennisi
University of Catania

KLUWER ACADEMIC PUBLISHERS
BOSTON / DORDRECHT / LONDON

A C.I.P. Catalogue record for this book is available from the Library of Congress.

ISBN 0-7923-7643-9

Published by Kluwer Academic Publishers,
P.O. Box 17, 3300 AA Dordrecht, The Netherlands.

Sold and distributed in North, Central and South America
by Kluwer Academic Publishers,
101 Philip Drive, Norwell, MA 02061, U.S.A.

In all other countries, sold and distributed
by Kluwer Academic Publishers,
P.O. Box 322, 3300 AH Dordrecht, The Netherlands.

Printed on acid-free paper

To our families:

Michela and Francesca

Stefania, Francesco, and Valeria

CONTENTS

ACKNOWLEDGEMENTS

The authors wish to thank Massimo Alioto, Walter Aloisi and Rosario Mita for their help during the correction of the draft.

A special thank is due to Professor John Choma jr., a scientific leader in feedback theory and feedback amplifiers, for his encouragement and inspiration in the development of this book.

We would like to thank our families and parents for their endless support and interest in our careers.

Gaetano Palumbo
Salvatore Pennisi

PREFACE

Feedback circuits and their related properties have been extensively investigated since the early days of electronics. From the time scientific and industrial communities started talking about and working with active elements like vacuum tubes or transistors, until today, much literature and many scientific results have been published which reinforce the importance of feedback. Improved features have been implemented in integrated circuits, novel techniques of analysis have been proposed which deeply improve our understanding of the resulting layouts, and new design strategies have been developed to optimise performance. Nevertheless, the genuinely complex subject of feedback and its applications in analog electronics remain obscure even for the majority of graduate electronics students.

To this end, the main focus of this book will be to provide the reader with a real and deep understanding of feedback and feedback amplifiers. Whenever possible and without any loss of generality, a simple and intuitive approach will be used to derive simple and compact equations useful in pencil-and-paper design. Complex analytical derivations will be used only when necessary to elucidate fundamental relationships. Consequently, the contents of the book have been kept to a reasonably accessible level.

The book is written for use both by graduate and postgraduate students who are already familiar with electronic devices and circuits, and who want to extend their knowledge to cover all aspects of the analysis and design of analog feedback circuits/amplifiers. Although the material is presented in a formal and theoretical manner, much emphasis is devoted to a design perspective. Indeed, the book can become a valid reference for analog IC designers who wish to deal more deeply with feedback amplifier features and their related design strategies, which are often partially –or even incorrectly– presented in the open literature. For this purpose (and despite maturity of the subject), novel formalisms, approaches, and results are described in this book. For instance, a generic small-signal model applicable to a variety of different transistor types operating in the active region is introduced. A new comprehensive approach for the frequency compensation of two-stage and three-stage amplifiers is adopted. Novel and insightful results are reported for harmonic distortion in the frequency domain.

The outline of the text is as follows:

Chapter 1 provides a brief introduction to the operating principles of Bipolar and MOS transistors together with their small-signal models. This chapter is an invited contribution by Dr. Ginaluca Giustolisi.

A general small-signal model for transistors in the active region of operation is derived in Chapter 2. The resulting model helps the reader to acquire a uniform view of the designer's tasks and sidesteps the impractical distinction traditionally practised between Bipolar and MOS devices. This model is then thoroughly utilised in the rest of the chapter and the book itself. The three basic single-transistor configurations, which are the common-emitter, common-collector, common-base, for the bipolar transistor and common source, common-drain, common-gate for the MOS transistor, are subsequently revisited. General relationships, for both these active components, valid at low and high frequencies are accordingly developed.

Feedback is introduced in Chapter 3. Feedback features are discussed in detail with particular emphasis on achievable advantages (and corresponding disadvantages) from a circuit perspective. Moreover, after an overview of the numerous techniques proposed until now to analyse feedback circuits, the two techniques which are the most useful in the authors' opinion are presented together with Blackman's theorem which is concerned only with the impedance level change due to feedback. Both techniques, namely the Rosenstark and the Choma methods, lead to exact results, but provide information only from a behavioural and approximated point of view. In fact, it is also demonstrated that these methods can bestow a deeper understanding of feedback properties.

Chapter 4 analyses the frequency and step response of transfer functions characterised by different combinations of poles (and zeros) that are commonly found in real practice. From this starting point, useful definitions will be given to help designers derive fundamental relations which ensure closed-loop stability with adequate margins.

Frequency compensation is a fundamental step in feedback amplifier design. Chapter 5 gives a classification of the most commonly employed compensation techniques. The traditional approaches such as dominant-pole compensation and Miller compensation are presented in detail with emphasis being devoted not only to the theoretical viewpoint, but also to a strong design perspective. Improved zero compensation techniques, which allow the frequency response of the resulting amplifier to be optimised, are then presented. In addition, the nested Miller approaches, which are becoming more and more important given the trend to reduce power supply, are also included..

Chapter 6 combines the knowledge introduced in the previous three chapters. The fundamental feedback amplifier architectures (Series-Shunt, Shunt-Series, Shunt-Shunt, and Series-Series topologies) are discussed assuming they are made with the general transistor introduced previously. Then practical applications are given for the two analysis and frequency compensation approaches.

Chapter 7 focuses on harmonic distortion in feedback amplifiers. Static non-linearity is analysed in a theoretical and exact manner. Moreover, the study of distortion versus frequency is carried out in a simple fashion and the results applied to the main frequency compensation techniques. We avoid traditional approaches such as the Volterra or Wiener series, which are computationally heavy, by exploiting considerations deriving from frequency compensation, which are mandatory in feedback amplifiers.

Chapter 8 deals with noise performance. Methods of analysis are illustrated and practical considerations and approximations which arise in real amplifiers, are included.

Chapter 9 looks at some examples taken from modern microelectronics which locally involve feedback or which are used in feedback configurations. The objective of the chapter is not only to show further practical examples, but also to outline some typical features inherent to these selected circuits. Thus a dual goal is achieved: acquiring more knowledge about the items treated in previous chapters, and gaining greater insight into some of the properties exhibited by well-known and useful circuits, strongly related to the topic of the book.

Lastly the Appendix summarises useful results related to the analysis of transfer functions of RC networks.

Chapter 1

INTRODUCTION TO DEVICE MODELING
Gianluca Giustolisi

This chapter will deal with the operation and modeling of semiconductor devices in order to give the reader a basis for understanding, in a simple and efficient manner, the operation of the main building blocks of microelectronics.

1.1 DOPED SILICON

A semiconductor is a crystal lattice structure with free electrons and/or free holes or, which is the same, with negative and/or positive carriers. The most common semiconductor is silicon which, having a valence of four, allows its atoms to share four free electrons with neighboring atoms thus forming the covalent bonds of the crystal lattice.

In intrinsic silicon, thermal agitation can endow a few electrons with enough energy to escape their bonds. In the same way, they leave an equal number of holes in the crystal lattice that can be viewed as free charges with an opposite sign. At room temperature, we have $1.5 \cdot 10^{10}$ carriers of each type per cm^3. This quantity is referred to as n_i and is a function of temperature as it doubles for every 11 °C increase in temperature [1]-[2].

This intrinsic quantity of free charges is not sufficient for the building of microelectronic devices and must be increased by doping the intrinsic silicon. This means adding negative or positive free charges to the pure material. Several doping materials can be used to increase free charges. Specifically, when doping pure silicon with a pentavalent material (that is, doping with atoms of an element having a valence of five) we have almost one extra free electron that can be used to conduct current for every one atom of impurity. Likewise, doping the pure silicon with atoms having a valence of three, gives us almost one free hole for every impurity atom. A pentavalent atom donates electrons to the intrinsic silicon and is known as a donor. In contrast, a trivalent atom accepts electrons and is known as an acceptor. Typical pentavalent impurities, also called n-type dopants, are arsenic, As, and phosphorus, P, while the most used trivalent impurity, also called p-type dopant, is boron, B. Silicon doped with a pentavalent impurity is said to be n-type silicon, while silicon doped with a trivalent impurity is called p-type silicon.

If we suppose that a concentration N_D of donor atoms (greater than the intrinsic carrier concentration, n_i) is used to dope the silicon, the concentration of free electrons in the n-type material, n_n, can be assumed as equal to

$$n_n = N_D \qquad (1.1)$$

In fact, this is an approximation, since some of the free electrons of the doping material recombine with the holes, but it is sufficient for as long as condition $N_D \gg n_i$ is true.

The fact that some free electrons recombine with holes, also reduces the concentration of holes in the n-type material, p_n, to

$$p_n = \frac{n_i^2}{N_D} \qquad (1.2)$$

Similarly, if we dope the silicon with a concentration N_A of acceptor atoms, the concentration of free holes in the p-type material, p_p, is equal to

$$p_p = N_A \qquad (1.3)$$

while the electron-hole recombination reduces the concentration of free electrons in the p-type material, n_p, to

$$n_p = \frac{n_i^2}{N_A} \qquad (1.4)$$

1.2 DIODES

A diode, or pn junction, is made by joining a p-type to an n-type material as in Fig. 1.1. The p-side terminal is called anode (A) while the n-side terminal is called cathode (K).

Note that the p-type section is denoted with p+, meaning that this side is doped more heavily (in the order of 10^{20} carriers/cm^3) than its n-type counterpart (in the order of 10^{15} carriers/cm^3), that is $N_A \gg N_D$. This is not a limitation since most pn junctions are built with one side more heavily doped than the other.

Close to the junction, free electrons on the n side are attracted by free positive charges on the p side so they diffuse across the junction and recombine with holes. Similarly, holes on the p side are attracted by electrons on the n side, diffuse across the junction and recombine with free electrons on the n side.

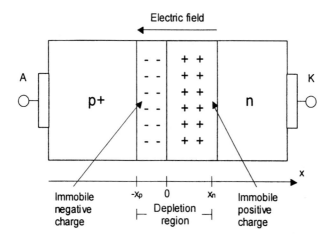

Fig. 1.1. pn junction.

This phenomenon leaves behind positive ions (or immobile positive charges) on the n side, and negative ions (or immobile negative charges) on the p side, thus creating a depletion region across the junction where no free carriers exist. Moreover, since charge neutrality obliges the total amount of charge on one side to be equal to the total amount of charge on the other, the width of the depletion region is greater on the more lightly doped side, that is, in our case where $N_A \gg N_D$, we have $x_n \gg x_p$.

Due to immobile charges, an electric field appears from the n side to the p side and generates the so-called built-in potential of the junction. This potential prevents further net movement of free charges across the junction under open circuit and steady-state conditions. It is given by [1]-[2]

$$\Phi_0 = U_T \ln\left(\frac{N_A N_D}{n_i^2}\right) \qquad (1.5)$$

U_T being the thermal voltage defined as

$$U_T = \frac{kT}{q} \tag{1.6}$$

where T is the temperature in degrees Kelvin (≈ 300 K at room temperature), k is the Boltzmann's constant ($1.38 \cdot 10^{-23}$ JK^{-1}) and q is the charge of an electron ($1.602 \cdot 10^{-19}$ C). At room temperature, U_T is approximately equal to 26 mV. Typical values of the built-in potential are around 0.9 V.

Under open circuit and steady-state conditions, it can be shown that the widths of depletion regions are given by the following equations

$$x_p = \left[\frac{2\varepsilon_{si}\varepsilon_0\Phi_0}{q} \frac{N_D}{N_A(N_A + N_D)} \right]^{1/2} \tag{1.7a}$$

$$x_n = \left[\frac{2\varepsilon_{si}\varepsilon_0\Phi_0}{q} \frac{N_A}{N_D(N_A + N_D)} \right]^{1/2} \tag{1.7b}$$

where ε_0 is the permittivity of free space ($8.854 \cdot 10^{-12}$ F/m) and ε_{si} is the relative permittivity of silicon (equal to 11.8).

Dividing (1.7a) by (1.7b) yields

$$\frac{x_n}{x_p} = \frac{N_A}{N_D} \tag{1.8}$$

which justifies the fact that x_n is greater than x_p if $N_A \gg N_D$. Moreover, under this condition, we can further simplify (1.7) in

$$x_n \approx \left(\frac{2\varepsilon_{si}\varepsilon_0\Phi_0}{qN_D} \right)^{1/2} \tag{1.9a}$$

$$x_p \approx \left(\frac{2\varepsilon_{si}\varepsilon_0\Phi_0 N_D}{qN_A^2} \right)^{1/2} \tag{1.9b}$$

The charge stored in the depletion region, per unit device area, is found by multiplying the width of the depleted area by the concentration of the immobile charge, which can be considered equal to q times the doping concentration. So for both the sides of the device we have

$$Q^- = qN_a x_p = \left(2q\varepsilon_{si}\varepsilon_0 \Phi_0 \frac{N_A N_D}{N_A + N_D} \right)^{1/2} \tag{1.10a}$$

$$Q^+ = qN_D x_n = \left(2q\varepsilon_{si}\varepsilon_0 \Phi_0 \frac{N_A N_D}{N_A + N_D} \right)^{1/2} \tag{1.10b}$$

Note that the charge stored on the n side equals the charge stored on the p side, as is expected due to the charge neutrality.

In the case of a more heavily doped side, as in our example where $N_A \gg N_D$, we can simplify (1.10) to

$$Q^+ = Q^- \approx \left[2q\varepsilon_{si}\varepsilon_0 \Phi_0 N_D \right]^{1/2} \tag{1.11}$$

1.2.1   Reverse Bias Condition

By grounding the anode and applying a voltage V_R to the cathode, we reverse-bias the device. Under such a condition the current flowing through the diode is mainly determined by the junction area and is independent of V_R. In many cases this current is considered negligible and the device is modeled as an open circuit. However, the device also has a charge stored in the junction that changes with the voltage applied and causes a capacitive effect, which cannot be ignored at high frequencies. The capacitive effect is due to the so-called junction capacitance.

Specifically, when the diode is reverse biased as in Fig. 1.2, free electrons on the n side are attracted by the positive potential V_R and leave behind positive immobile charges. Similarly, free holes in the p region move towards the anode leaving behind negative immobile charges. This means that the depletion region increases and that the built-in potential increases exactly by the amount of applied voltage, V_R.

Given that the built-in potential is increased by V_R, both the width and the charge of the depletion region can be found by substituting the term $\Phi_0 + V_R$ to Φ_0 in (1.7) and (1.10), respectively. In particular the charge stored results as

$$Q^+ = Q^- = \left[2q\varepsilon_{si}\varepsilon_0 (\Phi_0 + V_R) \frac{N_A N_D}{N_A + N_D} \right]^{1/2} \tag{1.12}$$

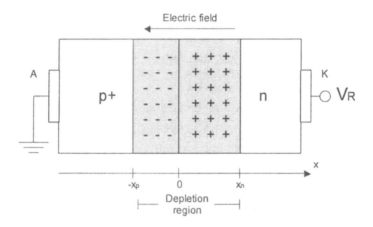

Fig. 1.2. Reverse-biased pn junction.

This charge denotes a non-linear charge-voltage characteristic of the device, modeled by a non-linear capacitor called a junction capacitance.

For small changes in the applied voltage around a bias value, V_R, the capacitor can be viewed as a small-signal capacitance, C_j, whose expression is found by differentiating[1] (1.12) with respect to V_R

$$C_j = \frac{dQ^+}{dV_R} = \frac{C_{j0}}{\sqrt{1 + \dfrac{V_R}{\Phi_0}}} \tag{1.13}$$

where

$$C_{j0} = \left(\frac{q\varepsilon_{si}\varepsilon_0}{2\Phi_0} \frac{N_A N_D}{N_A + N_D} \right)^{1/2} \tag{1.14}$$

is a capacitance per unit of area and depends only on the doping concentration.

1.2.2   Graded Junctions

All the above equations are valid in the case of abrupt junctions. For graded junctions, that is where the doping concentration changes smoothly

---

[1] All the derivatives are evaluated at the quiescent operating point.

from p to n, a better model for the charge can be described by changing the exponent in (1.12) as follows [4]

$$Q^+ = Q^- = \left[2q\varepsilon_{si}\varepsilon_0 (\Phi_0 + V_R) \frac{N_A N_D}{N_A + N_D} \right]^{1-m} \tag{1.15}$$

where m is a technology dependent parameter (typical m values are around 1/3).

In this case, the junction capacitance per unit of area turns into

$$C_j = \frac{dQ^+}{dV_R} = \frac{C_{j0}}{\left(1 + \frac{V_R}{\Phi_0}\right)^m} \tag{1.16}$$

where

$$C_{j0} = \frac{1-m}{\Phi_0^m} \left(2q\varepsilon_{si}\varepsilon_0 \frac{N_A N_D}{N_A + N_D} \right)^{1-m} \tag{1.17}$$

1.2.3 Forward Bias Condition

With reference to Fig. 1.3, by grounding the cathode and applying a voltage V_D to the anode, we forward-bias the device. Under this condition the built-in potential is reduced by the amount of voltage applied. Consequently, the width of the depletion region and the charge stored in the junction are reduced, too.

If V_D is large enough, the reduction in the potential barrier ensures the electrons in the n side and the holes in the p side are attracted by the anode and the cathode, respectively, thus crossing the junction. Once free charges cross the depletion region, they become minority carriers on the other side and a recombination process with majority carriers begins. This recombination reduces the minority carrier concentrations that assume a decreasing exponential profile. The concentration profile is responsible for the current flow near the junction, which is due to a diffusive phenomenon that is called diffusion current. On moving away from the junction, some current flow is given by the diffusion current and some is due to majority carriers that, coming from the terminals, replace those carriers recombined with minority carriers or diffused across the junction. This latter current is termed a drift current.

This process causes a current to flow through the diode that is exponentially related to voltage V_D as follows

$$I_D = A_D J_S \exp\left(\frac{V_D}{U_T}\right) \tag{1.18}$$

where A_D is the junction area and J_S the scale current density which is inversely proportional to the doping concentrations. The product $A_D J_S$ is often expressed in terms of a scale current and denoted as I_S.

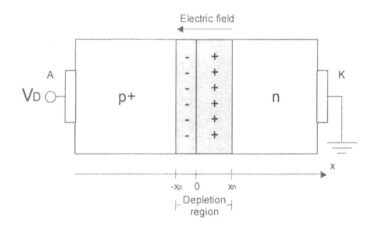

Fig. 1.3. Forward-biased pn junction.

As far as the charge stored in the device is concerned, we have two contributions under the forward bias condition. The first is given by the charge stored in the depletion region, Q_j, that can be evaluated by substituting $-V_D$ for V_R in (1.12), assuming there is an abrupt junction. In the same manner, this charge yields a small signal junction capacitance that can be expressed by (1.13) and (1.14). In any case, since this contribution is negligible, the junction capacitance is often modeled with a capacitive value of $2C_{j0}$ where C_{j0} is expressed by (1.14) or (1.17), depending on whether the junction is assumed to be abrupt or not.

The second contribution takes into account the charge due to minority carrier concentrations close to the junction that are responsible for the diffusion current. This component yields a diffusion capacitance, C_d, which is proportional to the current I_D as follows [1]-[2]

$$C_d = \tau_T \frac{I_D}{U_T} \tag{1.19}$$

where τ_T is a technology parameter known as the transit time of the diode.

The total capacitance, C_T, is the sum of the diffusion capacitance, C_d, and the junction capacitance, C_j, that is

$$C_T = C_d + C_j \approx C_d + 2C_{j0} \tag{1.20}$$

1.2.4   Diode Small Signal Model

In the case of reverse bias, the diode can be simply modeled with the junction capacitance defined by (1.13) and (1.14) or by (1.15) and (1.17), depending on whether the junction is abrupt or graded.

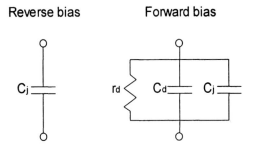

Fig. 1.4. Diode small signal models.

In the case of forward bias a small signal resistor, r_d, models the current-voltage relationship. Specifically, from (1.18) we have

$$r_d = \left(\frac{dI_D}{dV_D} \right)^{-1} = \left[\frac{A_D J_S}{U_T} \exp\left(\frac{V_D}{U_T} \right) \right]^{-1} = \frac{U_T}{I_D} \tag{1.21}$$

The capacitive contribution is taken into account by adding the capacitor C_T in (1.20) in parallel to r_d. Diode small signal models are depicted in Fig. 4.

1.3 MOS TRANSISTORS

Currently, Metal-Oxide-Semiconductor Field-Effect Transistors (MOSFETs or simply MOS transistors) are the most commonly used components in integrated circuit implementations since their characteristics make them more attractive than other devices such as, for example, BJTs.

Specifically, their simple realization and low cost, the possibility of having a complementary technology with the same characteristics for both complementary devices, their small geometry and, consequently, the feasibility of integrating a large number of devices in a small area, their infinite input resistance at the gate terminal and the faculty of building digital cells with no static dissipation, all motivate the great success of MOS transistors in modern technologies.

A simplified cross section of an n-channel MOS (n-MOS) transistor is shown in Fig. 1.5. It is built on a lightly doped p type substrate (p-) that separates two heavily doped n type regions (n+) called source and drain. A dielectric of silicon oxide and a polysilicon gate are grown over the separation region. The region below the oxide is the transistor channel and its length, that is the length that separates the source and the drain, is the channel length, denoted by L. In present MOS technologies the channel length is typically between 0.18 μm and 1 μm. In a p-channel MOS (p-MOS) all the regions are complementary doped.

There is no physical difference between the source and the drain as the device is symmetric, the notations source and drain only depend on the voltage applied. In an n-MOS the source is the terminal at the lower potential while, in a p-MOS, the source is the terminal at the higher potential.

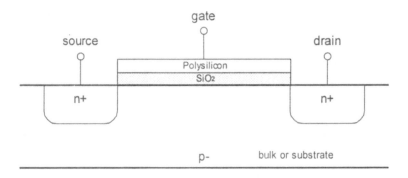

Fig. 1.5. Simplified cross section of an n- MOS transistor.

1.3.1   Basic Operation

To understand the basic operation of MOS transistors we shall analyze the behavior of an n-MOS depending on the voltages applied at its terminals.

If source, drain and substrate are grounded, the device works as a capacitor. Specifically, the gate and the substrate above the SiO_2 interface are two plates electrically insulated by the silicon oxide.

If we apply a negative voltage to the gate, negative charges will be stored in the polysilicon while positive charges will be attracted to the channel region thus increasing the channel doping to p+. This situation leads to an accumulated channel. Source and drain are electrically separated because they form two back-to-back diodes with the substrate. Even if we positively bias either the source or the drain, only a negligible current (the leakage current) will flow from the biased n+ regions to the substrate.

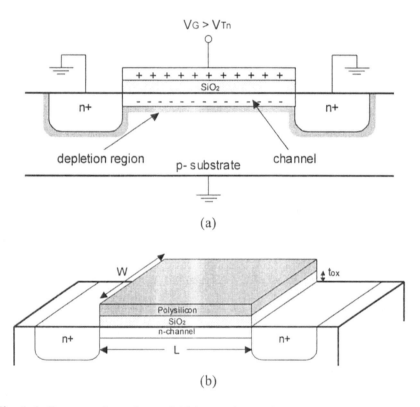

(a)

(b)

Fig. 1.6. Cross section of an n-MOS transistor when the channel is present:
a) bidimensional, b) tridimensional.

By applying a positive voltage to the gate, positive charges will be stored in the gate. Below the silicon oxide, if the gate voltage is small, positive free charges of the p- substrate will be repelled from the surface thus depleting the channel area. A further increase in the gate voltage leads to negative free charges being attracted to the channel that thereby becomes an n region. In this condition the channel is said to be inverted.

The gate-source voltage for which the concentration of electrons under the gate equals the concentration of holes in the p- substrate far from the gate is said to be the transistor threshold voltage, V_{Tn}.

At a first approximation, if the gate-source voltage, V_{GS}, is below the threshold voltage, no current can exist between the source and the drain and the transistor is said to be in the cutoff region. In contrast, if the gate-source voltage is greater than the threshold voltage, an n channel joins the drain and the source and a current can flow between these two electrically connected regions.

Actually, for gate voltages around V_{Tn}, the charge does not change abruptly and a small amount of current can flow even for small negative values of $V_{GS} - V_{Tn}$. This condition is termed weak inversion and the transistor is said to work in subthreshold region.

When the channel is present, as in Fig. 1.6, the accumulated negative charge is proportional to the gate source voltage and depends on the oxide thickness, t_{ox}, since the transistor works as a capacitor. Specifically, the charge density of electrons in the channel is given by [1]-[2]

$$Q_n = C_{ox}(V_{GS} - V_{Tn})$$ (1.22)

where C_{ox} is the gate capacitance per unit area defined as

$$C_{ox} = \frac{\varepsilon_{ox}\varepsilon_0}{t_{ox}}$$ (1.23)

and ε_{ox} is the relative permittivity of the SiO_2 (ε_{ox} is approximately 3.9)

The total capacitance and the total charge are obtained by multiplying both the equations (1.22) and (1.23) by the device area, as follows

$$C_{gs} = WLC_{ox}$$ (1.24a)

$$Q_{T-n} = WLC_{ox}(V_{GS} - V_{Tn})$$ (1.24b)

1.3.2   Triode or Linear Region

Increasing the drain voltage, V_D, causes a current to flow from the drain to the source through the channel. A drain voltage different from zero will modify the charge density but for small V_D the channel charge will not change appreciably and can be expressed by (1.22) again. Under this condition, the device operates as a resistor of length L, width W with a permittivity proportional to Q_n. Therefore, the relationship between voltage V_{DS} and the drain-source current, I_D, can be written as [7]

$$I_D = \mu_n Q_n \frac{W}{L} V_{DS} \tag{1.25}$$

where μ_n is the mobility of electrons near the silicon surface.

Substituting (1.22) in (1.25) yields

$$I_D = \mu_n C_{ox} \frac{W}{L} (V_{GS} - V_{Tn}) V_{DS} \tag{1.26}$$

Larger drain voltages modify the charge density profile in the channel. Specifically, referring to Fig. 1.7, we can express the channel charge density as a function of channel length. For $x = 0$, that is, close to the source, (1.22) holds, while for $x = L$, that is, close to the drain, we have

$$Q_n(L) = C_{ox}(V_{GD} - V_{Tn}) \tag{1.27}$$

Assuming a linear profile, the charge density has the following expression

$$Q_n(x) = \frac{Q_n(L) - Q_n(0)}{L} x + Q_n(0) \tag{1.28}$$

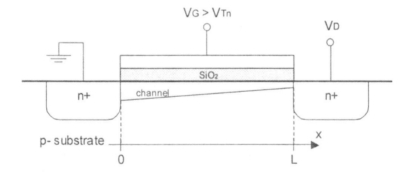

Fig. 1.7. MOS transistor channel for large V_D.

The current can be expressed in a form similar to (1.25) but with a different charge expression. If the charge density profile is linear, the average charge density can be used instead. The average charge density results in

$$\overline{Q}_n = \frac{Q_n(L) + Q_n(0)}{2} = C_{ox}\left(V_{GS} - V_{Tn} - \frac{V_{DS}}{2}\right) \tag{1.29}$$

and substituting this value in (1.25) leads to

$$I_D = \mu_n \overline{Q}_n \frac{W}{L} V_{DS} = \mu_n C_{ox} \frac{W}{L} \left(V_{GS} - V_{Tn} - \frac{V_{DS}}{2} \right) V_{DS} \qquad (1.30)$$

The current I_D is linearly related to V_{GS} and has a quadratic dependence on V_{DS}. Under this condition the device is said to operate in triode or linear region. Note also that (1.30) is reduced to (1.26) for small values of V_{DS}.

1.3.3   Saturation or Active Region

A further increase of V_D, can lead to the condition of a gate-drain voltage equal to V_{Tn}. In this case the charge density close to the drain, $Q_n(L)$, becomes zero and current I_D reaches its maximum value. This condition is shown in Fig. 1.8.

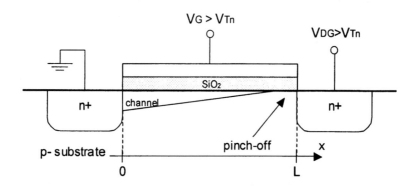

Fig. 1.8. MOS transistor channel for $V_{DG} > V_{Tn}$.

At a first approximation, the current does not change over this point with V_{DS} since the charge concentration in the channel remains constant and the electron carriers are velocity saturated. Under this condition the transistor is said to work in saturation or linear region.

Denoting V_{DSsat} as the drain source voltage when the charge density $Q_n(L)$ becomes zero, we can find an equivalent relationship that expresses the pinch-off condition by substituting $V_{DG} = V_{DS} - V_{GS}$ into $V_{DG} > V_{Tn}$. Specifically, we get

$$V_{DS} > V_{DSsat} \qquad (1.31)$$

where

$$V_{DSsat} = V_{GS} - V_{Tn} \tag{1.32}$$

Substituting the value $V_{DS} = V_{DSsat}$ defined in (1.32) into (1.30) gives the current expression in the pinch-off case and results as

$$I_D = \frac{\mu_n C_{ox}}{2} \frac{W}{L} (V_{GS} - V_{Tn})^2 \tag{1.33}$$

As mentioned above, (1.33) is valid at a first approximation. In fact, increasing V_D yields an increase in the pinch-off region as well as a decrease in channel length. This effect is commonly known as channel length modulation. To take this effect into account, a corrective term is used to complete (1.33) which becomes

$$I_D = \frac{\mu_n C_{ox}}{2} \frac{W}{L} (V_{GS} - V_{Tn})^2 [1 + \lambda (V_{DS} - V_{DSsat})] \tag{1.34}$$

The parameter λ is referred to as the channel length modulation factor and, at a first approximation, it is inversely proportional to the channel length, L.

1.3.4 Body Effect

All the equations derived above were based on the assumption that the source and the substrate (or the bulk) were connected together. Although this is a rather common condition, in general the voltage of these two terminals can be different. In this event a second order effect occurs commonly referred to as the *body effect* [6]. A different voltage between the source and the bulk is modeled as an increase in the threshold voltage, which assumes the following expression [6]-[8]

$$V_{Tn} = V_{Tn0} + \gamma \left(\sqrt{V_{SB} + 2|\phi_F|} - \sqrt{2|\phi_F|} \right) \tag{1.35}$$

with V_{SB} being the source-bulk voltage, V_{Tn0} the threshold voltage with zero V_{SB}, ϕ_F the Fermi potential of the substrate and γ a constant referred to as the body-effect constant. The Fermi potential is defined as [1]

$$\phi_F = -\frac{kT}{q} \ln \left(\frac{N_A}{n_i} \right) \tag{1.36}$$

while the value of γ depends on the substrate doping concentration as follows [1]

$$\gamma = \frac{\sqrt{2qN_A\varepsilon_{si}}}{C_{ox}} \qquad (1.37)$$

1.3.5   p-channel Transistors

For a p-channel transistor we can use the same equations derived in the previous sections, provided that a negative sign is placed in front of every voltage variable.

Therefore, V_{GS} becomes V_{SG}, V_{DS} becomes V_{SD}, V_{Tn} becomes $-V_{Tp}$, and so on. Note that in a p-MOS transistor the threshold voltage is negative. The condition for a p-MOS to be in saturation region is now $V_{SD} > V_{SG} + V_{Tp}$. Current equations (1.30) and (1.34) still hold but the current now flows from the source to the drain.

1.3.6   Saturation Region Small Signal Model

The low-frequency small signal model for a MOS transistor operating in the active region is shown in Fig. 1.9.

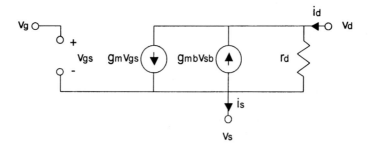

Fig. 1.9. Low-Frequency small signal model for a MOS transistor in active region.

The most important small signal component is the dependent current generator, $g_m v_{gs}$, whose transconductance, g_m, is defined as

$$g_m = \frac{\partial I_D}{\partial V_{GS}} = \mu_n C_{ox} \frac{W}{L}(V_{GS} - V_{Tn}) \qquad (1.38)$$

Solving (1.33) for $V_{GS} - V_{Tn}$ yields

$$V_{GS} - V_{Tn} = \sqrt{\frac{I_D}{\mu_n C_{ox} \frac{W}{2} \frac{W}{L}}} \tag{1.39}$$

and substituting this value in (1.38) we get the well-known expression for the transconductance

$$g_m = \sqrt{2\mu_n C_{ox} \frac{W}{L} I_D} \tag{1.40}$$

Another useful expression for g_m can be found by comparing (1.38) with (1.33) thus obtaining

$$g_m = \frac{2I_D}{V_{GS} - V_{Tn}} \tag{1.41}$$

The second dependent current source, $g_{mb}v_{sb}$, accounts for the body effect and its transconductance is defined as

$$g_{mb} = -\frac{\partial I_D}{\partial V_{SB}} = -\frac{\partial I_D}{\partial V_{Tn}} \frac{\partial V_{Tn}}{\partial V_{SB}} \tag{1.42}$$

The first derivative in (1.42) results as

$$\frac{\partial I_D}{\partial V_{Tn}} = -\mu_n C_{ox} \frac{W}{L} (V_{GS} - V_{Tn}) = -g_m \tag{1.43}$$

while the second one comes out by deriving (1.35) with respect to V_{SB}, thus yielding

$$\frac{\partial V_{Tn}}{\partial V_{SB}} = \frac{\partial}{\partial V_{SB}} \left[V_{Tn0} + \gamma \left(\sqrt{V_{SB} + 2|\phi_F|} - \sqrt{2|\phi_F|} \right) \right] = \frac{\gamma}{2\sqrt{V_{SB} + 2|\phi_F|}} \tag{1.44}$$

Therefore, substituting (1.43) and (1.44) in (1.42) we get

$$g_{mb} = \frac{\gamma g_m}{2\sqrt{V_{SB} + 2|\phi_F|}} \tag{1.45}$$

Note that this value is nonzero even if the quiescent value of V_{SB} equals zero. Specifically, the body effect arises only if a small signal, v_{sb}, is present between the source and the bulk terminals. In general g_{mb} is 0.1–0.2 times g_m and can be neglected in a non-detailed analysis.

The last model parameter is the resistor r_d, which takes into account the channel length modulation or, which is the same, the dependence of the drain current on V_{DS}. It is related to the large signal equations by

$$\frac{1}{r_d} = \frac{\partial I_D}{\partial V_{DS}} \tag{1.46}$$

Substituting in (1.46) the current expression in (1.34) results as

$$\frac{1}{r_d} = \frac{\partial I_D}{\partial V_{DS}} = \frac{\partial}{\partial V_{DS}} \frac{\mu_n C_{ox}}{2} \frac{W}{L} (V_{GS} - V_{Tn})^2 [1 + \lambda(V_{DS} - V_{DSsat})]$$

$$= \lambda \frac{\mu_n C_{ox}}{2} \frac{W}{L} (V_{GS} - V_{Tn})^2 \approx \lambda I_D \tag{1.47}$$

and finally

$$r_d \approx \frac{1}{\lambda I_D} \tag{1.48}$$

The high-frequency model of a MOS transistor, which includes the capacitive effects, is shown in Fig. 1.10.

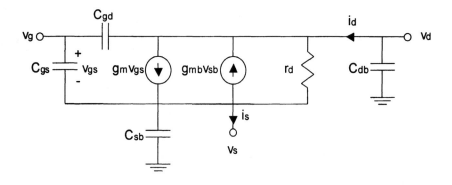

Fig. 1.10. High-frequency small signal model for a MOS transistor in active region.

Each capacitive contribution has its own physical meaning that can be understood by analyzing the detailed n-MOS cross section in Fig. 1.11.

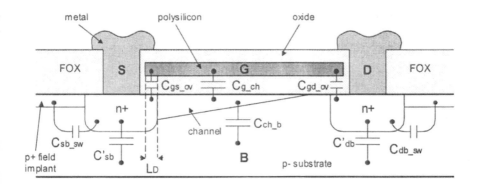

Fig. 1.11. Detailed n-MOS cross-section.

The most important capacitor is the gate-source capacitor whose value is given by two different terms. The first term takes into account the capacitive effect between the gate and the channel, which is electrically connected to the source. At a first approximation, the corresponding capacitor, C_{g_ch}, is a linear capacitor that depends on the oxide thickness as well as on the device area. It can be demonstrated that its value is approximately given by [4], [7]

$$C_{g\_ch} \approx \frac{2}{3} WL C_{ox} \tag{1.49}$$

The second term that contributes to the gate-source capacitance is given by the overlap that exists between the gate and the source n+ region. This overlap is unavoidable and results from the fact that during the fabrication process the doping element also spreads horizontally. Naming L_D the overlap diffusion length, the resulting parasitic capacitor, C_{gs_ov}, is given by

$$C_{gs\_ov} = WL_D C_{ox} \tag{1.50}$$

Hence, the capacitor C_{gs} in Fig. 1.9 is expressed by the sum of (1.49) and (1.50), that is

$$C_{gs} = WC_{ox} \left(\frac{2}{3} L + L_D \right) \tag{1.51}$$

Since $L_D \ll L$ the capacitor C_{gs} is mainly determined by the gate-channel capacitance and the overlap effect can be neglected in many cases.

The same boundary effect that determines the gate-source overlap capacitance yields the gate-drain capacitance that is given by

$$C_{gd} = C_{gd\_ov} = WL_D C_{ox} \qquad (1.52)$$

This capacitor makes a strong contribution when the transistor is used as a voltage amplifier and a large voltage gain exists between the drain and the gate. In all other cases its contribution is negligible.

The second largest capacitor is the source-bulk capacitor, which can be split into three contributions all of them given by the depletion capacitances of reverse biased pn junctions. The first, C'_{sb}, takes into account the junction capacitance between the n+ source area and the bulk. Its expression is similar to (1.13) or (1.16) depending on whether the junction can be considered as abrupt or graded. Assuming a graded junction we have

$$C'_{sb} = \frac{A_s C_{j0}}{\left(1 + \dfrac{V_{SB}}{\Phi_0}\right)^m} = A_s C_{js} \qquad (1.53)$$

where A_s is the area of the source junction and C_{js} is defined as the source junction capacitance per unit area.

The second contribution is responsible for C_{ch_b} and takes into account the depletion region between the channel and the bulk. Even in this case we have an expression similar to (1.53) that is

$$C_{ch\_b} = \frac{A_{ch} C_{j0}}{\left(1 + \dfrac{V_{SB}}{\Phi_0}\right)^m} = A_{ch} C_{js} \qquad (1.54)$$

where A_{ch} is the area of the channel which can be evaluated as WL.

The third term is referred to as the source-bulk sidewall capacitance and is denoted as C_{sb_sw}. This capacitance is due to the presence of a highly p+ doped region (field implant) that exists under the thick field oxide (FOX) and prevents the leakage current from flowing between two adjacent transistors. The value of C_{sb_sw} can be particularly large if the field implant is heavily doped as in modern technologies. The expression of C_{sb_sw} is then

$$C_{sb\_sw} = \frac{P_s C_{j-sw0}}{\left(1 + \dfrac{V_{SB}}{\Phi_0}\right)^m} = P_s C_{js-sw} \tag{1.55}$$

where P_s is the perimeter of the source junction, excluding the side adjacent to the channel. Both C_{j-sw0} and C_{j-sw} are capacitances per unit length. Consequently, the source-bulk capacitance is given by the sum of (1.53), (1.54) and (1.55), that is

$$C_{sb} = \left(A_s + A_{ch}\right)C_{js} + P_s C_{js-sw} \tag{1.56}$$

The fourth capacitor in the model in Fig. 1.9 is the drain-bulk capacitor, C_{db}. This is similar to the source-bulk capacitance except for the fact that the channel does not make any contribution. Therefore equations similar to (1.53) and (1.55) can be written as follows

$$C'_{db} = \frac{A_d C_{j0}}{\left(1 + \dfrac{V_{DB}}{\Phi_0}\right)^m} = A_d C_{jd} \tag{1.57a}$$

$$C_{db\_sw} = \frac{P_d C_{j-sw0}}{\left(1 + \dfrac{V_{DB}}{\Phi_0}\right)^m} = P_d C_{jd-sw} \tag{1.57b}$$

and the drain-bulk capacitance results as

$$C_{db} = A_d C_{jd} + P_d C_{jd-sw} \tag{1.58}$$

1.3.7   Triode Region Small Signal Model

The low frequency small signal model for a MOS in triode region is a resistor whose value can be determined by deriving (1.39) with respect to V_{DS}, that is

$$\frac{1}{r_d} = \frac{\partial I_D}{\partial V_{DS}} = \mu_n C_{ox} \frac{W}{L}\left(V_{GS} - V_{Tn} - V_{DS}\right) \tag{1.59}$$

If V_{DS} is small (1.59) is often approximated by

$$\frac{1}{r_d} = \mu_n C_{ox} \frac{W}{L} \left(V_{GS} - V_{Tn} \right) \qquad (1.59)$$

The high-frequency model is not easy to determine because the channel is directly connected to both the source and drain resulting in a distributed RC network over the whole length of the device. Moreover, because of the capacitive nature of junction capacitances, the capacitive elements are highly non-linear. This is another factor making the model quite complicated for management by hand analysis.

A simplified model, which is quite accurate for small V_{DS}, can be obtained by evaluating the total channel charge contribution and by assuming half of this contribution to be referred to the source and half to the drain [7].

Specifically, since the total gate-channel capacitance is given by

$$C_{g\_ch} = WLC_{ox} \qquad (1.60)$$

gate-source and gate-drain capacitances can be modeled as

$$C_{gs} = C_{gd} = WC_{ox}\left(\frac{L}{2} + L_D\right) \qquad (1.61)$$

where the overlap contribution has been included.

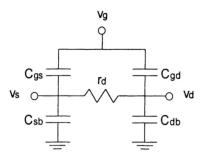

Fig. 1.12. High-frequency small signal model for a MOS transistor in triode region.

In the same way, the channel-bulk contribution is shared between the source and the drain thus yielding for C_{sb} and C_{db} the following

$$C_{sb} = \left(A_s + \frac{A_{ch}}{2} \right) C_{js} + P_s C_{js-sw} \tag{1.62a}$$

$$C_{db} = \left(A_d + \frac{A_{ch}}{2} \right) C_{jd} + P_d C_{jd-sw} \tag{1.62b}$$

The resulting high-frequency small signal model is depicted in Fig. 1.12

1.3.8   Cutoff Region Small Signal Model

In the cutoff region the resistance r_d is assumed to be infinite so the equivalent model is purely capacitive.

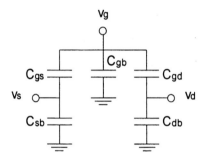

Fig. 1.13. High-frequency small signal model for a MOS transistor in cutoff region

Since the channel is not present, both C_{gd} and C_{gs} are due only to the overlap contribution, that is

$$C_{gs} = C_{gd} = WL_D C_{ox} \tag{1.63}$$

Source-bulk and drain-bulk capacitances are similar to those given in (1.62) with the difference that the channel does not make any contribution, that is

$$C_{db} = A_d C_{jd} + P_d C_{jd-sw} \tag{1.64a}$$

$$C_{sb} = A_s C_{js} + P_s C_{js-sw} \tag{1.64b}$$

The fact that no channel exists, generates a new capacitor, C_{gb}, which connects the gate and the bulk. Its value is given by the oxide capacitance multiplied by the device area, that is

$$C_{gb} = A_{ch}C_{ox} = WLC_{ox} \qquad\qquad (1.65)$$

The resulting small signal model is shown in Fig. 13.

1.3.9   Second Order Effects in MOSFET Modeling

The main second order effects that should be taken into account when determining a MOS large signal model are reported in this section. Their effects are always present but are especially prominent in short-channel devices and, often, cannot be ignored.

In the following we shall neglect the subscript n, which referred to n-MOS transistors.

1.3.9.1   Channel length reduction due to overlap

Referring to Fig. 1.11, we see that designed channel, L, is reduced due to the overlap. Assuming a symmetric device with equal overlap, L_D, at both the source and the drain, the amount of reduction is equal to $2L_D$, that is, the effective channel length, L_{eff}, is equal to

$$L_{eff} = L - 2L_D \qquad\qquad (1.66)$$

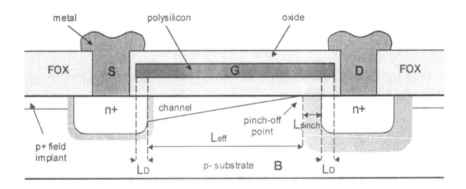

Fig. 1.14. Channel length modulation.

Obviously, the influence of the overlap is greater in short channel devices as it strongly affects the real channel. As a consequence, in all the previous equations, (1.66) should be used for the channel length.

A similar equation holds for the width, W, as well ($W_{eff} = W - 2W_D$). However, this effect is less frequent since minimum MOS widths are hardly chosen especially in analog designs. Thus we can assume $W \approx W_{eff}$.

1.3.9.2 Channel length modulation

The channel length modulation was discussed in previous sections and was modeled by the channel length modulation factor, λ, in (1.34). In this model, the pinch-off point was assumed to be close to the drain end.

A more effective modeling would take into consideration the fact that, in practice, the pinch-off point moves towards the source as V_{DS} increases due to the variation in the drain depletion region. As a consequence, the effective channel length, L_{eff}, is further reduced as shown in Fig. 1.14. Defining L_{pinch} as the distance between the drain end and the pinch-off point we get

$$L_{eff} = L - 2L_D - L_{pinch} \qquad (1.67)$$

The value of L_{pinch} is a function of V_{DS}, V_{DSsat} and the doping concentration of the channel. Substituting, for example, (1.67) in (1.33) we observe that, due to a shorter channel, the drain current increases with V_{DS}. Obviously, this effect is particularly evident in short-channel devices [8].

1.3.9.3 Mobility reduction due to vertical electric field

As known, the mobility, μ, relates the electrical field, E, to the drift velocity of carriers, v_d, as [1]-[2]

$$v_d = \mu E \qquad (1.68)$$

In our previous model we assumed the mobility to be a constant. Actually the value of this parameter depends on several physical factors, the most important of which is related to the carrier-scattering mechanisms.

The carrier scattering in the channel is greatly influenced by the vertical electric field induced by the gate voltage. Consequently mobility changes with V_{GS}. A semi-empirical equation used to model the mobility reduction due to vertical fields is [6]

$$\mu_s = \frac{\mu_0}{1 + \vartheta(V_{GS} - V_T)} \qquad (1.69)$$

where μ_s is now the new mobility (or better the new surface mobility), μ_0 is the mobility in the case of low fields and ϑ is the mobility degradation factor whose value can be related to oxide thickness as $2.3/t_{ox}$ (nm) [6].

It can be shown that this effect can be modeled as a series resistance, R_S, in the source of the MOS where

$$R_S \approx \frac{\vartheta}{\mu_0 C_{ox} W / L_{eff}}$$ (1.70)

1.3.9.4   Mobility reduction due to lateral electric field

Mobility is further reduced due to the high lateral electric field. Since, at a first approximation, the electric field is proportional to V_{DS}/L_{eff}, this effect is more pronounced in short-channel devices.

The linear relationship (1.68), which relates the drift velocity to the electric field, no longer holds for high fields because the mobility strongly depends on the field itself and decreases as the field increases. Specifically, at high electric fields, the drift velocity of carriers deviates from the linear dependency in (1.68) and even saturates. To account for this physical phenomenon, the mobility, μ_s, in (1.69) is corrected as follows [6]

$$\mu_{eff} = \frac{\mu_s}{1 + \dfrac{\mu_s}{v_{max}} \dfrac{V_{DS}}{L_{eff}}}$$ (1.71)

where μ_{eff} is the effective mobility, V_{DS}/L_{eff} represents the lateral field and the term v_{max} is the maximum drift velocity of the carriers. A typical value of v_{max} is in the order of a 10^5 m/s.

This velocity limitation can be responsible for the saturation in MOS transistors since a MOS can enter the active region before V_{DS} reaches the value of $V_{GS} - V_T$. Consequently, (1.32) must be adjusted to account for the carrier saturation velocity.

1.3.9.5   Drain Induced Barrier Lowering (DIBL)

This effect is due to the strong lateral electric field and affects the threshold voltage. The principal model assumes the channel is created by the gate voltage only. Actually, a strong lateral field from the drain can also help to attract electrons towards the surface. Strictly speaking, the drain voltage influences the surface charge and helps the gate voltage to form the channel. This effect is modeled with a reduction in the threshold voltage (that is, a barrier lowering) and is also modeled by modifying (1.35) as [6], [8]

$$V_T = V_{T0} + \gamma \left(\sqrt{V_{SB} + 2|\phi_F|} - \sqrt{2|\phi_F|} \right) - \sigma_D V_{DS}$$ (1.72)

where σ_D is a corrective factor responsible for the dependence of the threshold voltage on V_{DS}.

Also the DIBL is more pronounced in short-channel devices and its main effect is a further reduction in the output resistance.

1.3.9.6 Threshold voltage dependency on transistor dimensions

As transistor dimensions are reduced, the fringing field at border edges can also affect the threshold voltage [8].

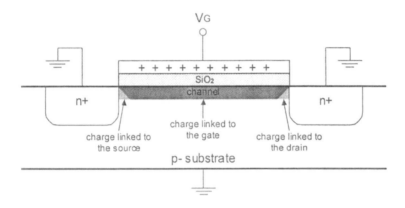

Fig. 1.15. Border effects in MOS transistors.

Referring to Fig. 1.15 and without entering into a detailed physical explanation, applying a voltage V_G to the gate creates a channel. However, due to border effects, only charges in the darker trapezoidal area are linked to the gate voltage. The threshold voltage definition in (1.35) refers all the charges in the rectangular area below the silicon to the gate voltage, as in Fig. 1.6. Since the threshold voltage depends on the channel charge linked to the gate voltage, it is apparent that the previous model overestimates the value of V_T. This border effect is not critical in long-channel devices, but in a short-channel transistor it can be significant.

To model this phenomenon, the threshold voltage in (1.35) is modified in

$$V_T = V_{T0} - \gamma\sqrt{2|\phi_F|} + F_s\gamma\sqrt{V_{SB} + 2|\phi_F|} \tag{1.73}$$

where F_s is a corrective factor that represents the ratio between the trapezoidal and the rectangular areas used to model the channel. As a consequence, V_T is less than its original value in (1.35).

In a similar way, the threshold voltage depends on transistor width if this dimension becomes comparable to the edge effect regions, that is, in narrow-channel (i.e., with short width) devices.

In this case, after applying a voltage to the gate, border effects deplete a wider region thus increasing the threshold voltage. This effect is modeled by adding the term $F_n(V_{SB} + 2|\phi_F|)$ to the original V_T in (1.35). F_n is a corrective factor that approaches zero in the case of wide channels.

Taking into account (1.72) and (1.73) the final form for the threshold voltage becomes

$$V_T = V_{T0} - \gamma\sqrt{2|\phi_F|} + F_s\gamma\sqrt{V_{SB} + 2|\phi_F|} - \sigma_D V_{DS} + F_n\left(V_{SB} + 2|\phi_F|\right) \quad (1.74)$$

1.3.9.7   Hot carrier effects

High lateral electric fields can generate high velocity carriers also called hot carriers. In short-channel devices, due to their high velocity, electron-hole pairs can be generated in the channel by impact ionization and avalanching. As a consequence, in n-MOS, a current of holes can flow from the drain to the substrate. This effect can be viewed as a finite resistance that connects the drain to the substrate and can result in a major limitation when realizing high impedance cascode structures.

Moreover, some hot carriers with enough energy can tunnel the gate oxide thus causing either a dc gate current or, if trapped in the oxide, a threshold voltage alteration. This latter phenomenon can drastically limit the long-term reliability of MOS transistors.

A further hot carrier effect is the so-called punch-through. It happens when the depletion regions of source and drain are so close each other that hot carriers with enough energy can overcome the short-channel region thus causing a current that is no longer limited by the drift equations. It is as if the channel were no longer present in the device and both source and drain areas were connected together. This phenomenon is limited by increasing the substrate doping which consequently limits the depletion region extensions. This effect not only lowers drain impedance but can also cause transistor breakdown.

1.3.10   Sub-threshold Region

In our previous modeling we assumed that no conduction could exist if the gate voltage is below the threshold. Actually, when the gate voltage is increased over the threshold, there is not an abrupt transition from the cut-off region to any of the conducting region. Specifically, a small drain current can flow at gate voltages a few millivolts below V_T. In this condition, the device is said to operate in sub-threshold or in weak inversion region and the current has an exponential relationship with the voltage applied.

Assuming long channel devices, the drain current is well expressed by [9]

$$I_D = I_{D0} \frac{W}{L} \exp\left(\frac{V_G}{nU_T}\right)\left[\exp\left(-\frac{V_S}{U_T}\right) - \exp\left(-\frac{V_D}{U_T}\right)\right] \tag{1.75}$$

where all voltages are referred to the substrate. In (1.75), I_{D0} is a characteristic current whose typical value is around 20 nA and n is a slope factor whose value is about 1.5. Specifically, n depends on the surface depletion capacitance, C_d, as

$$n = 1 + \frac{C_d}{C_{ox}} \tag{1.76}$$

The weak inversion condition is expressed as

$$I_D \leq \frac{n-1}{e^2} \mu C_{ox} \frac{W}{L} U_T^2 \tag{1.77}$$

Often, to be sure that a transistor is operating in sub-threshold mode, (1.77) is roughly approximated to

$$I_D \ll \mu C_{ox} \frac{W}{L} U_T^2 \tag{1.78}$$

Assuming $V_{DS} > 75$ mV and the source short-circuited to the bulk, the drain current exponential relationship can be simplified into

$$I_D \approx I_{D0} \frac{W}{L} \exp\left(\frac{V_{GS}}{nU_T}\right) \tag{1.79}$$

Despite the exponential relationship, the drain current is very small and so is the transconductance g_m (equal to I_D/nU_T). Therefore the device is very slow. Moreover, it should be noted that transistor matching is very poor (due to threshold voltage variations) and that minimum size devices are normally avoided when working in weak inversion. This leads to large parasitic capacitances that further decrease device speed.

1.4 BIPOLAR-JUNCTION TRANSISTORS

Bipolar transistors or BJT were widespread until the end of seventies when MOS technology started to become popular thanks to the fact that a larger number of transistors could be put together in a single integrated circuit. Today, silicon-based integrated circuits are mainly fabricated using CMOS processes, with a small slice devoted to bipolar technologies.

With respect to MOS, bipolar transistors have the advantage of a larger transconductance factor, g_m, and a larger output resistance, r_c, so they exhibit better performance in terms of current driving capability and achievable voltage gain. Unfortunately, unlike MOS transistors, their base resistance is finite and drastically decreases at high frequencies.

Recent bipolar transistors have a unity-gain frequency in the order of 20-50 GHz. Since MOS unity-gain frequencies are one order of magnitude less, BJTs are mainly used in RF or high-speed digital integrated circuits.

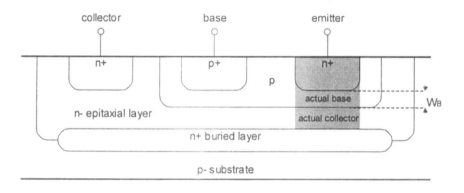

Fig. 1.16. Bipolar-Junction Transistor cross section.

A typical simplified BJT cross-section is shown in Fig. 1.16, where the so-called npn vertical transistor is depicted. It can be seen as two back-to-back diodes because it is made up of two n-regions separated by a p-region called base. The actual base region is the gray p-region in the figure whose width, W_B, is small with respect to the other proportions and in modern bipolar processes is between 0.5-0.8 μm. This region has a medium doping concentration, in the order of 10^{17} carriers/cm^3. The emitter is the heavily doped n+ region in the figure. It has a width of a few μm and its doping concentration is in the order of 10^{21} carriers/cm^3. Finally, the actual collector region is the gray n- epitaxial layer in the figure. The collector doping concentration is in the order of 10^{15} carriers/cm^3. To reduce the resistive path that connects the actual collector region to the collector contact, a heavily doped buried layer is grown below the device. The gray area represents the region where the so-called transistor effect takes place and is the actual npn transistor. Since this area extends vertically, the transistor is said to be vertical. Finally, note that, unlike for MOSFETs, the transistor is not symmetric.

1.4.1   Basic Operation

To understand the basic operation of a bipolar transistor let us consider the simplified scheme in Fig. 1.17 where the emitter terminal is connected to ground. The base-emitter junction acts as a diode and a current flows if the junction is forward biased. In such a situation, that is $V_{BE} > 0$, a current of majority carriers (holes in this case) flows from the base terminal across the base-emitter junction. Meantime, a current of electrons flows from the emitter across the base-emitter junction and enters the base thus diffusing towards the base-collector junction. Due to the different doping levels electrons that diffuse into the base are much more than just holes that diffuse into the emitter [1]-[3].

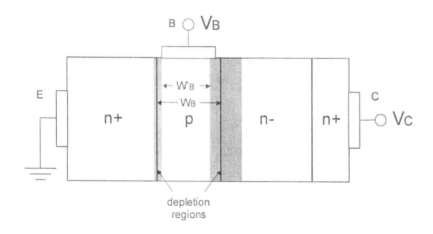

Fig. 1.17. Simplified scheme of a BJT.

If V_C is larger than 0.2-0.3 V, the excess of electrons in the base is subject to a negative electric field imposed by the collector voltage. When those electrons appear at the base-collector junction, they are pushed into the collector region. Since the base width, W_B, is small, electrons coming from the emitter do not have the possibility to recombine with holes in the base and almost all are pushed into the collector.

In such a situation, the small base current is mainly determined by holes while electrons coming from the emitter mainly determine the large collector current. Consequently, the emitter current is the sum of those two contributions. Under this condition the transistor is said to operate in the active region.

The collector current, I_C, is caused by the base-emitter voltage and, as for a diode, it has an exponential relationship that is

$$I_C = A_E J_{CS} \exp\left(\frac{V_{BE}}{U_T}\right) \tag{1.80}$$

where A_E is the emitter area and J_{CS} is a constant term that represents a current density and is inversely proportional to the base width, W_B, and its doping concentration. The product $A_E J_{CS}$ is often expressed in terms of the current scale factor I_{CS}.

In addition, the base current is exponentially related to the base-emitter voltage and has an expression similar to (1.80). Consequently, at a first approximation, the ratio between the collector and the base current is constant and independent of both voltages and currents. This ratio is commonly referred to as β_F, that is

$$\beta_F = \frac{I_C}{I_B} \tag{1.81}$$

Due to the small amount of base current with respect to the large collector current, the value of β_F is typically between 50 and 200.

The ratio between the collector current and the emitter current, I_E, is denoted with α_F and results as

$$\alpha_F = \frac{I_C}{I_E} \tag{1.82}$$

Since $I_E = I_C + I_B$, the constant α_F can be expressed in terms of β_F, that is

$$\alpha_F = \frac{\beta_F}{\beta_F + 1} \approx 1 \tag{1.83}$$

which is close to unity for high values of β_F.

1.4.2   Early Effect or Base Width Modulation

In (1.80) the collector current is independent of the collector voltage. However, this is true only at a first order approximation since the dependence in fact exists. Referring to Fig. 1.17, we note that the effective base width, W_B', that should be used for evaluating J_{CS}, is different from the designed base width, W_B, due to the presence of two depletion regions. The base-emitter depletion region is caused by a forward biasing. Therefore, it is small and almost independent of the voltage applied. In contrast, reverse biasing creates the base-collector depletion region, which, consequently, is larger and strongly depends on the voltage applied. Specifically, the collector voltage modulates the base-collector depletion region, thus decreasing and influencing the effective base width, W_B'.

To take this effect into account, a corrective term is introduced in (1.80) that becomes [6]

$$I_C = A_E J_{CS} \exp\left(\frac{V_{BE}}{U_T}\right)\left(1 + \frac{V_{CE}}{V_A}\right) \tag{1.84}$$

where the constant V_A is commonly referred to as the *Early Voltage* and has a typical value between 50 and 100 V. In most of the applications, especially for large signal analysis, this effect is negligible.

1.4.3 Saturation Region

When the collector-emitter voltage, V_{CE}, approaches the value of about 0.2-0.3 V, commonly referred to as V_{CEsat}, the base-collector junction becomes forward biased. In such a situation, holes from the base start to diffuse into the collector, and the collector current is no longer related to the base current by (1.81). Specifically, the base-emitter junction behaves like a diode whose current I_B exponentially depends on V_{BE} while the base-collector junction behaves like a voltage source whose value is set to V_{CEsat}.

1.4.4 Charge Stored in the Active Region

In the active region, the base-emitter junction is forward-biased so, like in a forward-biased diode, the charge stored across this junction is derived from two contributions. The first takes into account the charge in the small depletion region, Q_{jbe}, and can be modeled by substituting $-V_{BE}$ to V_R in (1.12). This charge leads to a junction capacitance whose value can be approximated by $2C_{j0be}A_E$ where C_{j0be} has an expression similar to (1.14) or (1.17) depending on whether the junction is assumed to be abrupt or not.

The second contribution is given by the minority carrier concentration in both the base and the emitter. As in a forward-biased diode, this contribution leads to a diffusion capacitance, C_{dbe}, expressed by

$$C_{dbe} = \tau_b \frac{I_C}{U_T} \tag{1.85}$$

where τ_b is a technology parameter commonly referred to as the base-transit-time constant. The total base-emitter capacitance, C_{be}, is then given by

$$C_{be} = 2C_{j0be} + C_{dbe} \approx C_{dbe} \tag{1.86}$$

In contrast, since the base-collector junction is reverse-biased, its relative charge is stored in the depletion region. Assuming a graded junction, the corresponding capacitance, C_{bc}, is expressed by an equation similar to (1.16) multiplied by the effective area of the base-collector interface, A_{BC}, that is

$$C_{bc} = \frac{A_{BC}C_{j0bc}}{\left(1 + \dfrac{V_{CB}}{\Phi_{c0}}\right)^m} \tag{1.87}$$

1.4.5   Active Region Small Signal Model

The most commonly used small signal model is the hybrid-π model. This model is similar to the small signal model for a MOSFET in saturation region and is shown in Fig. 1.18. Note that for historical reasons, small signal subscripts that refer to the base-emitter components, i.e. *be*, are substituted by the Greek letter π and small signal subscripts that refers to the base-collector components, i.e. *bc*, are substituted by the Greek letter μ.

As in the MOSFET case, the most important parameter is the dependent current generator, $g_m v_\pi$, whose transconductance, g_m, is defined as

$$g_m = \frac{\partial I_C}{\partial V_{BE}} = \frac{\partial}{\partial V_{BE}} I_{CS} \exp\left(\frac{V_{BE}}{U_T}\right) = \frac{I_C}{U_T} \tag{1.88}$$

The small signal resistance, r_π, is defined as

$$r_\pi = \left(\frac{\partial I_B}{\partial V_{BE}}\right)^{-1} = \left(\frac{1}{\beta_F} \frac{\partial I_C}{\partial V_{BE}}\right)^{-1} = \frac{\beta_F}{g_m} \tag{1.89}$$

The resistor r_c models the Early effect and is defined as

$$r_c = \left(\frac{\partial I_C}{\partial V_{CE}}\right)^{-1} = \left[\frac{\partial}{\partial V_{CE}} I_{CS} \exp\left(\frac{V_{BE}}{U_T}\right)\left(1 + \frac{V_{CE}}{V_A}\right)\right]^{-1} \approx \frac{V_A}{I_C} \tag{1.90}$$

All the above dc small-signal components are intrinsic terms of any BJT since they depend on the npn junction itself.

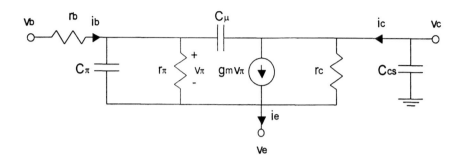

Fig. 1.18. Small signal model for a BJT in active region.

The model in Fig. 1.18 also includes a base resistance, r_b, which comes out in a real implementation. Specifically, r_b, models the resistive path that exists between the effective transistor base region (i.e. the gray area in Fig. 1.16) and the base contact (i.e. the p+ doped region). This path presents a small ohmic resistance of a few tens or hundreds of ohm. With respect to r_π, r_b has a small value and, in low frequency operations, it can be neglected since the base-emitter voltage is practically equal to v_π. In high-frequency circuits (i.e. in RF applications), part of the base current flows across C_π thus reducing the effective impedance in the base-emitter branch. Because of the presence of r_b, v_π can be significantly different from the base-emitter voltage applied thus considerably affecting transistor properties. In practice, r_b cannot be neglected if a high-speed circuit is being analyzed or designed.

Note that there is also an ohmic resistance in series with the actual collector (whose value is lowered by the n+ buried layer) but its presence is not as crucial as the base resistance is.

As far as the capacitive contribution is concerned, we have two main intrinsic capacitors, C_π and C_μ, as well as capacitor, C_{cs}, which exists in integrated implementations only.

Specifically, capacitor C_π, is the base-emitter capacitor and is expressed by (1.86), while C_μ, which represents the base-collector capacitive contribution, is expressed by (1.87). Due to their nature, C_μ is at least one order of magnitude smaller than C_π and, in several cases, is neglected. However, its contribution becomes significant when a high gain exists between the base and the collector.

Capacitor C_{cs}, comes out from the reverse biased pn region realized by the collector-substrate junction. This capacitor is quite large and is modeled by the following expression

$$C_{cs} = \frac{A_{CS} C_{j0cs}}{\left(1 + \dfrac{V_{CS}}{\Phi_{s0}}\right)^m} \tag{1.91}$$

where A_{CS} is the effective collector-substrate area, C_{j0cs} is the collector-to-substrate capacitance per unit area and Φ_{s0} is the built-in potential for the collector-substrate junction.

REFERENCES

[1]   S. Sze, *Physics of Semiconductor Devices*, John Wiley & Sons, 1981.

[2]   R. Muller, T. Kamins, *Device Electronics for Integrated Circuits*, John Wiley & Sons, 1986.

[3]   R. Gregorian, G. Temes, *Analog MOS Integrated Circuits for signal processing*, John Wiley & Sons, 1986.

[4]   P. Antognetti, G. Massobrio, *Semiconductor Device Modeling with SPICE*, McGraw Hill, 1988

[5]   P. Gray, R. Meyer, *Analysis and Design of Analog Integrated Circuits (third edition)*, John Wiley & Sons, 1993.

[6]   K. Laker, W. Sansen, *Design of Analog Integrated Circuits and Systems*, Mc Graw-Hill, 1994.

[7]   D. Johns, K. Martin, *Analog Integrated Circuit Design*, John Wiley & Sons, 1997.

[8]   Y. Cheng, C. Hu, *MOSFET Modeling & BSIM3 User's Guide*, Kluwer Academic Publishers, 1999

[9]   E. Vittoz, J. Fellrath, "CMOS Analog Integrated Circuits Based on Weak Inversion Operation," *IEEE J. Solid-State Circuits*, vol. SC-12, pp. 224-231, June 1977.

Chapter 2

SINGLE TRANSISTOR CONFIGURATIONS

2.1 THE GENERIC ACTIVE COMPONENT

In order to ensure that all analytical results derived herein are applicable to feedback configurations realised with both BJT and MOSFET technologies –including heterostructure bipolar transistor (HBT) and III-V compound metal-semiconductor field effect transistor (MESFET) technologies– we introduce the generic transistor component, whose circuit symbol and low-frequency small-signal model are shown in Fig. 2.1a-b and Fig. 2.1c, respectively [PC981].

This device is identified by four terminals denoted as X, Y, Z and B. Specifically, X, Y, and Z respectively representing the emitter, base, and collector terminals for BJTs (and HBTs) or the source, gate, and drain of MOSFETs (and MESFETs). The fourth terminal B represents the substrate or bulk node, it is almost always biased at a fixed potential and conducts a negligible static current. Symbols X, Y and Z were chosen to remind us of the functional equivalence between our generic device and the *negative second generation Current Conveyor* (CCII-) [SS70], [TLH90], [PPP99]. The ideal negative[1] CCII is a three-terminal device labelled by X, Y and Z (see Fig. 2.2) and is characterized by the following port relation

$$
\begin{bmatrix} v_x \\ i_y \\ i_z \end{bmatrix} = \begin{bmatrix} 0 & 1 & 0 \\ 0 & 0 & 0 \\ -1 & 0 & 0 \end{bmatrix} \begin{bmatrix} i_x \\ v_y \\ v_z \end{bmatrix}
\tag{2.1}
$$

---

[1] In a positive CCII the direction of the current flow at terminal Z has an opposite sign to a CCII-, i.e. $i_z = +i_x$.

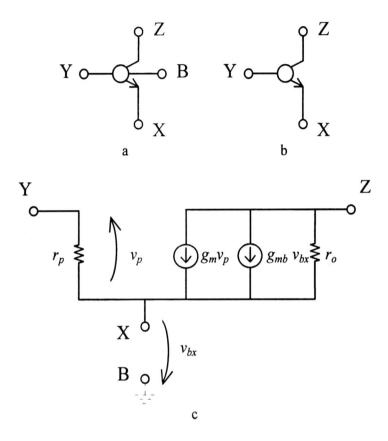

Fig. 2.1. Generic transistor used to represent BJT, HBT, MOSFET, and MESFET devices: Circuit symbol with the bulk terminal a), and without the bulk terminal b). Low-frequency small-signal model c).

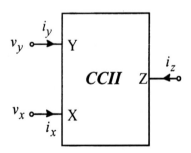

Fig. 2.2. Circuit symbol of a second-generation current conveyor.

The resistance at terminal Y is ideally infinite (no current flows through Y). The current flowing into terminal Z is a replica (but with opposite sign) of that flowing into terminal X. The voltage at terminal X is a replica of that applied to Y. The current in X can be supplied directly through terminal X itself, or can be produced by copying the voltage at terminal Y acting across an external impedance connected to terminal X.

As can be argued from Fig. 2.1b, the small-signal model of the generic transistor is derived by merging the BJT and MOSFET models. In order to further simplify the model, we could relate the bulk transconductance, g_{mb}, with transconductance g_m, via the bulk transconductance parameter, λ_b, (do not mistake this parameter for the MOS channel-length modulation parameter, λ)

$$g_{mb} = \lambda_b g_m \tag{2.2}$$

Parameter λ_b, is always much lower than 1. Its value is around 0.2-0.3 for MOSFETs and is equal to zero for BJTs. Thus, since resistance, r_p, is infinite for MOSFETs, the following relationship *always* holds

$$\frac{\lambda_b}{r_p} = 0 \tag{2.3}$$

An important parameter, not shown in the model, is the current gain, β, equal to

$$\beta = g_m r_p \tag{2.4}$$

Its value is usually in the range of 50 to 200 for BJTs, and, of course, is infinitely large for MOSFETs.

Resistance r_o is for any kind of transistor large enough to justify its neglect in discrete realisations employing discrete (load) resistors, while it must be often considered in IC applications. Hence, when appropriate and for the sake of completeness, we will include r_o in our analytical derivations.

2.2 AC SCHEMATIC DIAGRAM AND LINEAR ANALYSIS

In this book, we will be mostly interested in the small-signal properties of (feedback) configurations rather than their bias details. Thus, to simplify our description and analysis, and to focus only on the performance of interest, we will regularly make use of the *AC schematic diagram*, i.e., a circuit diagram in which biasing details are not shown.

Although small-signal analysis could be performed directly on the original schematic diagram (experienced designers do this), in this chapter we will not follow this procedure. For the sake of clarity (and for educational purposes) we will derive from the small-signal equivalent circuit from an AC schematic by replacing each transistor with its small-signal model and evaluating the circuit parameters. Later in the book, we can briefly refer to the results obtained below without requiring further investigations.

The small-signal equivalent circuit is obtained by linearising the circuit around its operating point. Figure 2.3 depicts a linear two-port network, whose output is terminated by a load resistance R_L. The output signals, voltage v_{out} or current i_{RL}, are generated in response to an input port signal whose equivalent voltage and resistance are respectively, v_{in} and R_S. Thus, a voltage and a current gain can be defined between v_{out} and v_{in}, and i_{RL} and i_{in}.

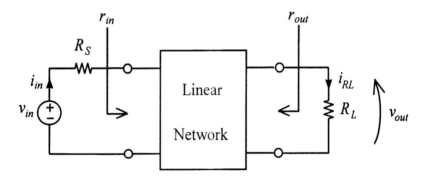

Fig. 2.3. Linear network with indicated input/output equivalent resistances.

Important parameters of this system are the input and output resistances seen at the input and output ports, which are responsible for a signal loss at the input and output when coupled with resistances R_S and R_L. We will denote these two resistances with r_{in} and r_{out}, respectively. The use of lowercase variables and subscripts is to remind us that these are small-signal resistances. Note that in general they depend on the load and source resistances, respectively.

In the rest of the chapter we will analyse the four basic single-transistor configurations. Since these schemes are covered in many electronic circuit textbooks, we will assume that the reader is familiar with their complete analysis (including biasing techniques, AC coupling, etc). Here we will recall only their salient small-signal features.

2.3 COMMON X (EMITTER/SOURCE) CONFIGURATION

Fig. 2.4a depicts the AC schematic diagram of the *common X* (CX) configuration. This circuit is used to provide voltage and current gain. The figure also includes resistance R_Y, which would not be strictly necessary, to render the analysis a more general one. It can account for the distributed base resistance of bipolar devices or the internal resistance of the signal generator. The small-signal circuit is illustrated in Fig. 2.4b. By defining the input and the output resistances of the common X configuration as illustrated in Fig. 2.4a, we can simply see by inspection that these are r_p and r_o, respectively.

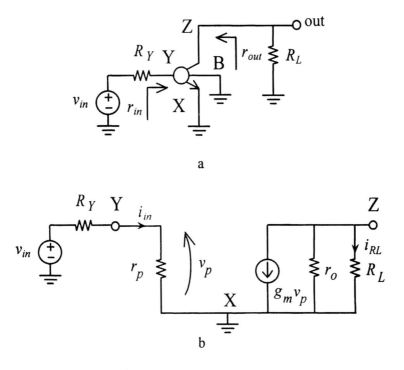

Fig. 2.4. Common X configuration: AC schematic diagram a), small-signal circuit b).

Voltage gain between terminal Y and Z can be evaluated considering that v_p equals v_y, and taking into account the input coupling between R_Y and r_p leading to the overall voltage gain

$$A_v = \frac{v_{out}}{v_{in}} = \frac{v_z}{v_{in}} = -\frac{r_p}{R_Y + r_p} g_m r_o \| R_L \tag{2.5}$$

The voltage gain is inverting, and neglecting the input loss, is reduced to $g_m R_L$ when the device output resistance, r_o, is much higher than the load resistance, R_L. Conversely, it achieves its maximum value, $g_m r_o$, which depends only on bias conditions and technological parameters, when the load resistance is much higher than the transistor output resistance. Also for this reason, quantity $g_m r_o$ is said to be the *intrinsic voltage gain* of the configuration.

The current gain is defined as the ratio between the current flowing into the Y terminal and that flowing in the load resistor to ground. It has a finite value only for bipolar devices and is strictly related to parameter β in (2.4).

$$A_i = \frac{i_{RL}}{i_{in}} = \frac{i_{RL}}{i_y} = -\beta \frac{r_o}{r_o + R_L} \tag{2.6}$$

Its magnitude decreases to a value lower than β when the output resistance cannot be considered much higher than the load resistance

2.4 COMMON X WITH DEGENERATIVE RESISTANCE R_X

The common X configuration with a local resistive feedback R_X is shown in Fig. 2.5a.

Although the main concern of this book is feedback, for the moment we will not consider the effects of resistance R_X from a feedback point of view, but analyse it directly from the small-signal circuit illustrated in Fig. 2.5b. To find the input resistance of the CX configuration we apply the KCL at terminal X

$$\frac{v_x}{R_X} + \frac{v_x - v_z}{r_o} - \frac{v_p}{r_p} - g_m v_p + g_{mb} v_x = 0 \tag{2.7}$$

By considering that $v_z = R_L \left(\dfrac{v_p}{r_p} - \dfrac{v_x}{R_X} \right)$ and solving (2.7) for v_x we get

$$v_x = \frac{R_X}{r_p} \frac{1 + g_m r_p + \dfrac{R_L}{r_o}}{1 + R_X \left(\lambda_b g_m + \dfrac{1}{r_o} \right) + \dfrac{R_L}{r_o}} v_p \tag{2.8}$$

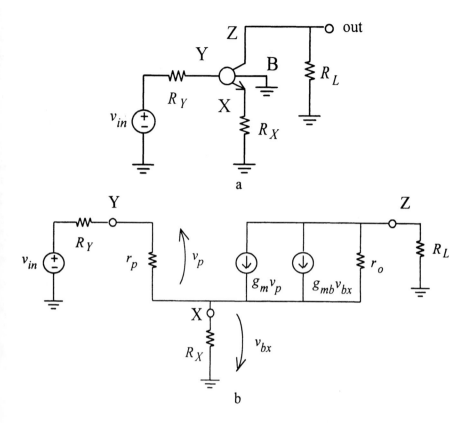

Fig. 2.5. Common X configuration with degenerative resistance: AC
schematic diagram a), small-signal circuit b).

Finally, the input resistance, r_{in}, can be expressed in terms of v_x/v_p

$$r_{in} = \frac{v_y}{i_y} = \frac{v_p + v_x}{i_y} = r_p + r_p \frac{v_x}{v_p} \qquad (2.9a)$$

Observing that r_{in} has an infinite value for MOS devices, we can avoid
considering the term which accounts for the bulk transconductance in (2.8).
Thus, (2.9a) can be usefully rewritten as

$$r_{in} = r_p + \left(1 + \frac{g_m r_p}{1 + \dfrac{R_L}{r_o}}\right)\left[(r_o + R_L)\| R_X\right] \qquad (2.9b)$$

Assuming R_L and R_X are lower than r_o, the input resistance r_{in} is accurately approximated by the well-known relation

$$r_{in} \approx r_p + (1 + \beta)R_X \tag{2.9c}$$

The equivalent output resistance is evaluated by setting the input node to ground. Hence, the small-signal circuit in Fig. 2.6 results.

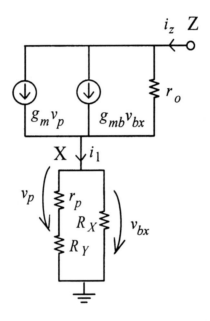

Fig. 2.6. Circuit for the evaluation of the output resistance in common X configuration with degenerative resistance.

By applying a test voltage, v_z, at terminal Z, we note that i_1 equals i_z, so v_z can be expressed as

$$v_z = (r_p + R_Y)\|R_X i_z + r_o\left[i_z - g_m\left(v_p + \lambda_b v_{bx}\right)\right] \tag{2.10a}$$

Since

$$v_{bx} = -(r_p + R_Y)\|R_X i_z \tag{2.10b}$$

and

$$v_p = -\frac{r_p}{r_p + R_Y}(r_p + R_Y)\|R_X\, i_z \tag{2.10c}$$

the output resistance is

$$r_{out} = r_z|_{v_Y=0} = \left.\frac{v_z}{i_z}\right|_{v_y=0} =$$

$$= r_o + \left[1 + g_m\left(\frac{r_p}{r_p + R_Y} + \lambda_b\right)r_o\right](r_p + R_Y)\|R_X \tag{2.11a}$$

It is apparent that resistance r_{out} has increased compared to the pure CX configuration. In particular, for MOS transistors the expression simplifies to

$$r_{out} = r_o + [1 + g_m(1 + \lambda_b)r_o]R_X \tag{2.11b}$$

Note that the body effect contributes to the output resistance growth. For bipolar devices we find the well-known result given in (2.11c), evaluated assuming resistance R_Y lower than r_p

$$r_{out} = r_o + (1 + g_m r_o)r_p\|R_X \tag{2.11c}$$

which tends to $r_p + (1 + \beta)r_o$ if R_X is greater than r_p.

To determine the voltage and current gain of the CX configuration, we make use of the *Norton* equivalent at the output of the circuit, as shown in Fig. 2.7a. The expression of the Norton resistance is exactly the same as given in (2.11). Then we only have to calculate the short-circuit output current. We prefer to perform this calculation in two steps:

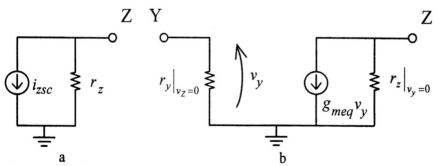

Fig. 2.7. Models of common X configuration with degenerative resistance: Output Norton equivalent a) and alternative small-signal model b).

First we evaluate the voltage partition between v_{in} and v_y

$$\left.\frac{v_y}{v_{in}}\right|_{v_z=0} = \left.\frac{r_y}{R_Y+r_y}\right|_{v_z=0} \tag{2.12}$$

Since we must set the node Z to ground, this is equivalent to setting R_L to zero. Using this condition in (2.9b) we obtain

$$\left. r_y\right|_{v_z=0} = r_p + \left(1+g_m r_p\right)r_o \| R_x \tag{2.13}$$

Then we evaluate the equivalent transconductance, g_{meq} by considering the short-circuit current flowing *into* terminal Z, which is expressed by

$$i_{zsc} = \left.i_z\right|_{v_z=0} = g_m v_p - \left(g_{mb} + \frac{1}{r_o}\right)v_x =$$

$$= \left(g_m - \frac{R_X}{r_p}\frac{1+g_m r_p}{\frac{1}{\lambda_b g_m}\| r_o + R_X}\right)v_p \approx g_m\left(1 - \frac{\lambda_b g_m R_X}{1+\lambda_b g_m R_X}\right)v_p \tag{2.14}$$

Thus, the equivalent transconductance is found to be

$$g_{meq} = \left.\frac{i_{zsc}}{v_p + v_x}\right|_{v_z=0} \approx \frac{g_m}{1+\lambda_b g_m R_X}\frac{1}{1+\frac{r_p}{R_X}\frac{1+g_m r_p}{1+\lambda_b g_m R_X + \frac{R_X}{r_o}}} \tag{2.15a}$$

where (2.8) with $R_L=0$ has been used.
Compared with the simple CX configuration, the degeneration reduces the transconductance value, although it provides higher input and output resistances.

    The above expression is further simplified if it is possible to neglect resistance r_o. Moreover, for bipolar transistors, or neglecting the body effect in MOS transistors, we get the well-known result

$$g_{meq} \approx \frac{g_m}{1+g_m R_X} \tag{2.15b}$$

which is lastly approximated by $1/R_X$, if $g_m R_X >> 1$. Observe, as a special case, that if R_X equals $1/g_m$, g_{meq} is approximated by $g_m/2$.

Finally, the voltage gain is

$$A_v = \frac{v_z}{v_{in}} = -i_{zsc} r_Z = -\left.\frac{r_y}{r_y + R_Y}\right|_{v_z=0} g_{meq} r_z \| R_L \qquad (2.16)$$

The above discussion is helpful in developing a simple model of the configuration considered, which is similar to the pure CX configuration, without explicitly including resistance R_X. The model is shown in Fig. 2.7b, with its parameters already defined in (2.11), (2.13), and (2.15). It can be used to simplify the analysis of complex circuits including, as building blocks, CX stages with degeneration. In particular, it provides exact voltage and current gains and output resistance. Note, however, that the input resistance r_y is evaluated with the output short-circuited. It (slightly) differs from the actual input resistance, which depends instead on R_L. Of course, r_{in} is very well approximated by $\left. r_y \right|_{v_z=0}$ if the condition $R_L < r_o$ is met.

The current gain is related to the equivalent parameter, $\beta_{eq} = \left. g_{meq} r_y \right|_{v_z=0}$, which is nearly equal to β. It is further reduced by the output coupling loss, so that

$$A_i = \frac{i_{RL}}{i_y} = -\left. g_{meq} r_y \right|_{v_z=0} \frac{r_z}{r_z + R_L} \approx -g_m r_p \frac{r_z}{r_z + R_L} \qquad (2.17)$$

Except for the lower loss at the output, it is similar to that of CX without the degenerative resistance.

2.5 COMMON Y (BASE/GATE)

The *common Y* (CY) configuration is generally exploited for its current following capability. In this configuration the X terminal is the input and Z is the output. Figure 2.8a depicts a general common Y topology where resistance R_Y, which is not strictly necessary, is again included to generalise the analysis.

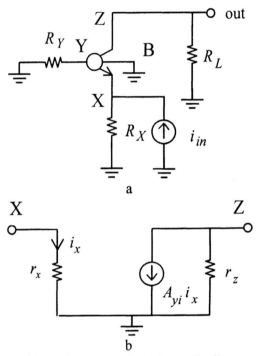

Fig. 2.8. Common Y configuration: AC schematic diagram a), alternative small-signal model b).

The input resistance, r_{in}, seen by input generator, i_{in}, in Fig. 2.8a, can be found by replacing the transistor with its model and writing the KCL at node X (we can simplify calculation by ideally removing resistance R_X and considering its effect later)

$$i_x - \frac{(v_x - v_z)}{r_o} + \frac{v_p}{r_p} + g_m v_p - g_{mb} v_x = 0 \qquad (2.18)$$

Since

$$v_p = -\frac{r_p}{r_p + R_Y} v_x \tag{2.19a}$$

and

$$v_z = R_L \left(i_x + i_p \right) \tag{2.19b}$$

remembering relationship (2.2), we obtain

$$r_{in} = \frac{v_x}{i_x} \| R_X =$$

$$= \frac{\left(r_p + R_Y \right)\left(1 + \dfrac{R_L}{r_o} \right)}{1 + \left(1 + \lambda_b \right)g_m r_p + \lambda_b g_m R_Y + \dfrac{r_p + R_Y + R_L}{r_o} + \dfrac{R_L R_Y}{r_o r_p}} \| R_X \tag{2.20a}$$

The above relation can be simplified if condition $R_Y < r_p$ is verified. In such a case we can distinguish two often encountered sub-cases

1) $R_L < r_o$ (simple load)

$$r_{in} = \frac{1}{\left(1 + \lambda_b \right)g_m} \| R_X \qquad \text{for MOSFETs} \tag{2.20b}$$

$$r_{in} = \frac{r_p}{1 + \beta} \| R_X \qquad \text{for BJTs} \tag{2.20c}$$

2) $R_L > r_o$ (e.g., cascode load)

$$r_{in} = \frac{R_L}{\left(1 + \lambda_b \right)g_m r_o} \| R_X \qquad \text{for MOSFETs} \tag{2.20d}$$

$$r_{in} = \frac{R_L r_p}{\beta r_o + R_L} \| R_X \qquad \text{for BJTs} \tag{2.20e}$$

The output resistance r_z is the same as in the previous configuration and is given by (2.11).

As for the CX configuration with degeneration, we develop an alternative model useful for simplifying gain calculation in more complex circuits. Again, due to the current-out nature of the configuration, it is convenient to represent the circuit at its output node with its Norton equivalent circuit. Figure 2.8b illustrates this model. The output resistance is r_z given in (2.11), and the input resistance (r_x evaluated with the output shorted to ground) can be derived directly by setting $R_L = 0$ in (2.20a) and can be rewritten as

$$r_x = \frac{v_x}{i_x}\bigg|_{v_z=0} = \frac{r_p + R_Y}{1+(1+\lambda_b)g_m r_p + \lambda_b g_m R_Y}\bigg\| r_o \qquad (2.20f)$$

The *inner* current gain, $A_{yi,}$ between the current flowing into the grounded terminal Z and the current i_x is

$$A_{yi} = \frac{i_z}{i_x}\bigg|_{v_z=0} = \frac{(1+\lambda_b)g_m r_p + \dfrac{R_Y + r_p}{r_o}}{(1+\lambda_b)g_m r_p + \dfrac{R_Y + r_p}{r_o}+1} \approx \frac{(1+\lambda_b)g_m r_p}{(1+\lambda_b)g_m r_p + 1} \qquad (2.21)$$

As expected, its value is very close to one. If we assume that a load resistance, R_L, is connected to the output node, the current gain of the complete amplifier is given by

$$A_i = \frac{i_{RL}}{i_{in}} = \frac{R_X}{R_X + r_x} A_{yi} \frac{r_z}{R_L + r_z} \qquad (2.22)$$

and the voltage gain is simply

$$A_v = \frac{v_z}{v_x} = A_i \frac{r_z\|R_L}{R_X\|r_x} \qquad (2.23a)$$

Assuming an ideally unitary inner current gain and a load resistance, R_L, which is much lower than the equivalent out resistance, r_z, (2.23a) can be approximated as

$$A_v \approx \frac{R_L}{r_x} \qquad (2.23b)$$

2.6 COMMON Z (COLLECTOR/DRAIN)

The *common Z* amplifier, shown in Fig. 2.9a, is used as voltage follower (or as level shifter). Its input is located at the Y terminal and the output is at the X terminal. The usual topology of the common Z amplifier is implemented without resistance R_Z, but to develop general relationships we include this resistance and only in the final step do we simplify the equations by eventually setting its value to zero.

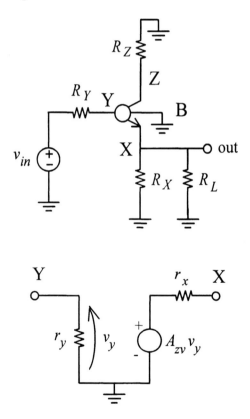

Fig. 2.9. Common Z configuration: AC schematic diagram a), alternative small-signal model b).

Except for the different designation of resistances at terminals X and Z, the calculation of the input resistance leads to exactly the same relationship as for the common X topology with a degenerative resistance. Here we report for the sake of completeness the expression given in (2.9a) rearranged for this case

$$r_{in} = \frac{v_y}{i_y} = r_p + \left(1 + \frac{g_m r_p}{1 + \frac{R_Z}{r_o}}\right)\left[(r_o + R_Z)\|(R_X\|R_L)\right] \tag{2.24}$$

The output resistance is similar (again, with different resistor names) to the input resistance of the common Y amplifier. Rearranging (2.20a) yields

$$r_{out} = \frac{\left(r_p + R_Y\right)\left(1 + \frac{R_Z}{r_o}\right)}{1 + (1 + \lambda_b)g_m r_p + \lambda_b g_m R_Y + \frac{r_p + R_Y + R_Z}{r_o} + \frac{R_Y R_Z}{r_o r_p}}\|R_X \tag{2.25}$$

Due to the voltage following function exhibited by this configuration, the alternative model, using which it is convenient to represent the circuit, is illustrated in Fig. 2.9b. In the model the output resistance is r_x given in (2.25), and the input resistance (r_y, evaluated with the output open) can be derived directly by making R_L infinitely large in (2.24). In this case, we are interested in finding the *Thévenin* equivalent of the circuit at the output, then the model exploits a voltage-controlled voltage source as a dependent source. To evaluate the equivalent voltage gain, A_{zv}, which represents the intrinsic voltage gain (i.e., without the source and load resistance) of the common Z amplifier, we use KCL at terminals X and Z

$$\frac{v_x}{R_X} + \frac{v_x - v_z}{r_o} - \frac{v_p}{r_p} - g_m v_p + g_{mb} v_x = 0 \tag{2.26a}$$

and

$$\frac{v_z}{R_Z} + \frac{v_x}{R_X} - \frac{v_p}{r_p} = 0 \tag{2.26b}$$

By substituting (2.26b) into (2.26a) and solving for v_x, we get

$$v_x = \frac{R_X}{r_p} \frac{1 + g_m r_p + \dfrac{R_Z}{r_o}}{1 + \dfrac{R_Z}{r_o} + R_X\left(\lambda_b g_m + \dfrac{1}{r_o}\right)} v_p \qquad (2.27)$$

By remembering that the value of r_p is finite only for the bipolar transistor we can set λ_b to zero. Using (2.27) and the relation $v_y = v_p + v_x$, after some manipulation we get

$$A_{zv} = \frac{v_x}{v_y} = \frac{g_m r_o + \dfrac{R_Z + r_o}{r_p}}{1 + (1 + \lambda_b)g_m r_o + \dfrac{R_Z + r_o}{r_p \| R_X}} \approx$$

$$\approx \frac{1 + g_m r_p}{1 + (1 + \lambda_b)g_m r_p + \dfrac{r_p}{R_X}} \qquad (2.28a)$$

As expected, the voltage gain is close to one. In particular, in bipolar technology the inaccuracy is due to the ratio between the bias resistance R_X and resistance r_p. Indeed, relationship (2.28a) becomes

$$A_{zv} = \frac{1 + \beta}{1 + \beta + \dfrac{r_p}{R_X}} \qquad \text{for BJTs} \qquad (2.28b)$$

On the other hand, in MOS implementations the loss is due to the bulk transconductance as shown below

$$A_{zv} = \frac{g_m R_X}{(1 + \lambda_b)g_m R_X + 1} \qquad \text{for MOSFETs} \qquad (2.28c)$$

The total voltage gain of the common Z amplifier including input and output loss is equal to

$$A_v = \frac{r_y}{R_Y + r_y} A_{zv} \frac{R_L}{R_L + r_X} \qquad (2.29)$$

The current gain is finite only for the bipolar technology, and it is given by

$$A_i = \frac{i_{RL}}{i_y} = A_v \frac{R_Y + r_y}{R_L}$$

(2.30a)

If we assume an ideally unitary inner voltage gain, A_{zv}, and a much higher input resistance than the source resistance, the current gain can be approximated as

$$A_i \approx \frac{r_y}{R_L}$$

(2.30b)

2.7 FREQUENCY RESPONSE OF SINGLE TRANSISTOR CONFIGURATIONS

To evaluate the frequency response of a circuit comprising the generic active component, we have to consider the transistor high-frequency small-signal model. This model is depicted in Fig. 2.10 and includes capacitors C_p, C_m, C_o and C_{xb}. It generally applies to all transistor types although, as discussed in Chapter 1, for each kind of device the capacitances have a different physical meaning.

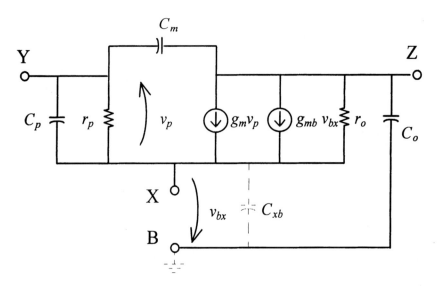

Fig. 2.10. High-frequency small-signal model of the generic transistor.

Of course, this is not the only possible model, more accurate models can be developed in particular to better approximate the behavior of the device for frequencies higher than its transition frequency. However, except for RF circuits, the model in Fig. 2.10 is sufficiently adequate to allow frequency behavior analysis and feedback amplifier compensation to be performed. Moreover, in the following we shall neglect the effect of the capacitance C_{xb} for simplicity, because it is not responsible for any substantial effect.

2.7.1 Common X Configuration

The AC schematic of the common X configuration is shown in Fig. 2.11a and the corresponding high-frequency small-signal model is depicted in Fig. 2.11b.

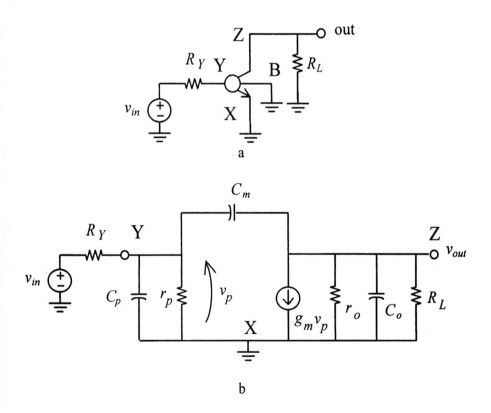

Fig. 2.11. Common X configuration: AC schematic diagram a), high-frequency small-signal circuit b).

The circuit voltage gain transfer function can be written as

$$A(s) = \frac{V_{out}(s)}{V_{in}(s)} = A_{vo} \frac{1 + b_1 s}{1 + a_1 s + a_2 s^2} \qquad (2.31)$$

where the DC voltage gain A_{vo} is given by (2.5). The coefficients a_1, a_2 and b_1 can be found by applying the procedure described in the Appendix or through a direct analysis of the circuit. They are given by

$$a_1 = r_p \| R_Y C_p + r_o \| R_L C_o + \left(r_p \| R_Y + r_o \| R_L + r_p \| R_Y g_m r_o \| R_L \right) C_m \qquad (2.32)$$

$$a_2 = r_p \| R_Y C_p r_o \| R_L (C_m + C_o) + r_p \| R_Y C_m r_o \| R_L C_o \qquad (2.33)$$

$$b_1 = -\frac{C_m}{g_m} \qquad (2.34)$$

We see that the contribution of C_m to coefficient a_1 (which approximately determines the first pole) can become dominant for high values of the voltage gain, $g_m r_o \| R_L$. The magnification of C_m is predicted by the well-known Miller theorem.

It is worth noting that coefficient b_1 related to the (positive) zero can be evaluated by inspection of Fig. 2.11b [MG91]. Indeed, at the complex frequency of the zero (i.e. $s = -1/b_1$) the output voltage must be zero and all the current of the controlled source, $g_m v_p$, flows through the capacitor C_m. Thus, after equating the two currents we get

$$g_m v_p = s C_m v_p \qquad (2.35)$$

which gives the frequency of the zero.

2.7.2 Common X with a Degenerative Resistance

Consider now the common X configuration with a degenerative resistance at node X, and its small-signal model shown in Fig. 2.12a and 2.12b, respectively.

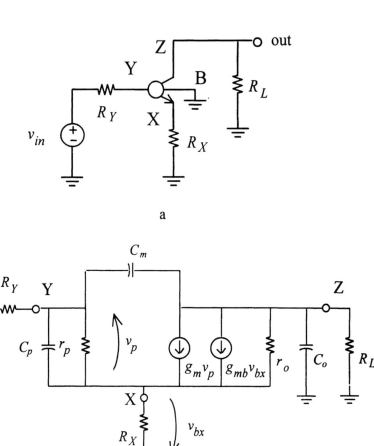

Fig. 2.12. Common X configuration with degenerative resistance: AC diagram a), small-signal circuit with parasitic capacitances b).

We again have a transfer function in the form of equation (2.31), where the DC voltage gain is now given by (2.16). By applying the method explained in the Appendix, the coefficients a_1, a_2 are given by

$$a_1 = r_{Cp}C_p + r_z\|R_L C_o + \left(r_y\|R_Y + r_z\|R_L + r_y\|R_Y \cdot g_m r_z\|R_L\right)C_m \qquad (2.36)$$

$$a_2 = r_{Cp}C_p\left(r_o + R_X\|R_Y\right)\|R_L\left(C_m + C_o\right) + r_y\|R_Y C_m r_z\|R_L C_o \qquad (2.37)$$

Alternatively, coefficients a_1 and a_2 can be obtained directly from (2.32) and (2.33) if we consider that we have to substitute r_o and r_p with r_z and r_y, respectively, to evaluate the equivalent resistance seen by capacitors C_m and C_o. Besides, the equivalent resistance seen by capacitance C_p, after neglecting transconductance g_{mb} and resistance r_o, is equal to

$$r_{Cp} = \frac{R_X + R_Y}{1 + R_X g_m} \| r_p \tag{2.38}$$

In fact, considering the small-signal circuit in Fig. 2.13, we get

$$i_p = -\frac{v_p + v_x}{R_Y} \tag{2.39}$$

and

$$v_x = R_X i_p + R_X g_m v_p \tag{2.40}$$

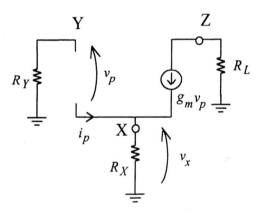

Fig. 2.13. Small-signal circuit to evaluate the equivalent resistance seen by capacitance C_p.

Coefficient b_1 can be obtained directly from (2.34) by substituting the transconductance g_m with its equivalent g_{meq} in the common X configuration with a degenerative resistance

$$b_1 = -\frac{C_m}{g_{meq}} \tag{2.41}$$

Similarly to the DC analysis, it would be useful to develop an equivalent model of the common X configuration with a degenerative resistance for the high-frequency analysis. The schematic of the model, which recalls that of a common X configuration without resistance R_X, is shown in Fig. 2.14.

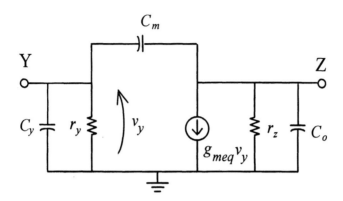

Fig. 2.14. High-frequency equivalent model of the common X configuration with a degenerative resistance.

It is apparent that to account for frequency behavior, we have introduced the new capacitor C_y, while the other two capacitors, C_m and C_o, are those intrinsic to the device model. After replacing the transistor and R_X in Fig. 2.12a with the above equivalent model, coefficients a_1 and a_2 of the transfer function become

$$a_1 = r_y \| R_Y C_y + r_z \| R_L C_o + \left(r_y \| R_Y + r_z \| R_L + r_y \| R_Y g_m r_z \| R_L \right) C_m \qquad (2.42)$$

$$a_2 = r_y \| R_Y C_p r_z \| R_L (C_m + C_o) + r_y \| R_Y C_m r_z \| R_L C_o \qquad (2.43)$$

Equating (2.42) and (2.43) with (2.36) and (2.38), respectively, under the assumption that r_o and hence r_z are much greater than R_L, we obtain the condition for the equivalence of the two models

$$r_y \| R_Y C_y = r_{Cp} C_p \qquad (2.44)$$

Hence, after some algebraic manipulation we get

$$C_y = \frac{\dfrac{R_X + R_Y}{1 + R_X g_m} \| r_p}{\left[r_p + (1 + r_p g_m) R_X \right] \| R_y} C_p \qquad C_p = \frac{(R_X + R_Y) r_p}{\left[r_p + (1 + r_p g_m) R_X \right] R_Y} C_p \qquad (2.45)$$

By neglecting r_p and R_X with respect to $g_m r_p R_X$ inside the square brackets of the denominator, the equivalent capacitance can be approximated to

$$C_y \approx \frac{C_p}{R_X \| R_Y g_m} \tag{2.46}$$

As expected, (2.46) shows a reduction of the equivalent capacitance seen at node Y. In reality, this *bootstrapping* effect is predicted by the well-known Miller theorem for those cases where the voltage gain across the capacitor tends to be unitary.

A more accurate evaluation of the zeros can be achieved by using the nodal admittance matrix as proposed in [R85]. The nodal admittance matrix can be written by inspection of the circuit in Fig. 2.12b in terms of the Laplace transform variable s. The device under consideration is identified by the four terminals X, Y, Z and B, where node B is connected to the reference node. Hence, the nodal admittance matrix can be defined as

$$Y = \begin{bmatrix} y_{yy} & y_{yx} & y_{yz} \\ y_{xy} & y_{xx} & y_{xz} \\ y_{zy} & y_{zx} & y_{zz} \end{bmatrix} \tag{2.47}$$

Evaluating the admittance terms inside the matrix [DK69] and neglecting resistance r_o and transconductance g_{mb}, we get

$$Y = \begin{bmatrix} G_y + g_p + s(C_p + C_m) & -(g_p + sC_p) & -sC_m \\ -(g_p + g_m + sC_p) & G_x + g_p + g_m + sC_p & 0 \\ g_m - sC_m & -g_m & G_L + sC_m \end{bmatrix} \tag{2.48}$$

where terms G_i represent the conductance (i.e. the inverse of the resistance R_i) of the i-th element. Of course, elements r_o and g_{mb} can be simply included in the matrix but at the cost of more complex expressions. Neglecting term g_p with respect to g_m, the nodal admittance matrix can be rewritten as

$$Y \approx \begin{bmatrix} G_y + g_p + s(C_p + C_m) & -(g_p + sC_p) & -sC_m \\ -(g_m + sC_p) & G_x + g_m + sC_p & 0 \\ g_m - sC_m & -g_m & G_L + sC_m \end{bmatrix} \tag{2.49}$$

As known from linear circuit theory, the transfer function numerator polynomial is an invariant characteristic of a circuit, while the denominator polynomial depends on the specific input and output terminals. In particular, it is analytically given by the determinant of the minor defined by the input and output node. Hence, for the common X configuration with degenerative resistance we get

$$\Delta_{yz} = \begin{vmatrix} -(g_m + sC_p) & G_x + g_m + sC_p \\ g_m - sC_m & -g_m \end{vmatrix} = C_p C_m s^2 + \frac{1 + g_m R_X}{R_X} C_m s - \frac{g_m}{R_X} \quad (2.50)$$

It is apparent that in the case where degenerative resistance R_X is equal to zero we find the same positive zero in (2.41), which also coincides with the approximated dominant zero obtained through (2.50).

2.7.3 Common Y and Common Z Configurations

Both the common Y and the common Z configurations can be represented by the transfer function

$$A(s) = A_o \frac{1 + b_1 s + b_2 s^2}{1 + a_1 s + a_2 s^2} \quad (2.51)$$

where the DC gain has already been derived in Sections 2.5 and 2.6. The coefficients a_1 and a_2 that identify the system poles are the same as in the common X configuration with the degenerative resistance. Indeed, the small-signal network is the same as the one in Fig. 2.12b after short-circuiting the voltage source. Regarding the numerator, we can again use the nodal admittance matrix (2.48). In particular, for the common Y configuration we get

$$\Delta_{xz} = -\begin{vmatrix} G_y + g_p + s(C_p + C_m) & -(g_p + sC_p) \\ g_m - sC_m & -g_m \end{vmatrix} \approx C_p C_m s^2 + g_m C_m s + \frac{g_m}{R_Y} \quad (2.52)$$

which shows two zeros that can be complex and conjugated.
For the common Z configuration the numerator of (2.51) is equal to

$$\Delta_{yx} = -\begin{vmatrix} -(g_m + sC_p) & 0 \\ g_m - sC_m & G_L + sC_m \end{vmatrix} = (g_m + sC_p)(G_L + sC_m) \quad (2.53)$$

which shows two real and negative zeros.

Chapter 3

FEEDBACK

Feedback, whether intentional or parasitic, is pervasive in all electronic circuits and systems. Usually, a feedback network includes a subcircuit that allows a fraction of the output signal to modify the effective input signal, in such a way as to produce a circuit response that can differ substantially from the response produced in the absence of such feedback. If the magnitude and the relative phase angle of the fed back signal decreases the magnitude of the signal applied to the input, the feedback is said to be *negative* or *degenerative*. Otherwise, the feedback is said to be *positive* or *regenerative*. Because negative feedback tends to produce stable[1] circuit responses, it is used in most applications. Note, however, that the parasitic feedback incurred by the energy storage elements associated with circuit layout, circuit packaging, and second order high frequency device phenomena often degrades an otherwise negative feedback circuit into a potentially regenerative or severely underdumped network. In contrast, positive feedback often enhances an inclination towards unstable behavior. This property proves useful when designing oscillators. Moreover, small amounts of positive feedback can be useful even in linear applications. For instance, to reduce component spreads (i.e., R_{max}/R_{min} and/or C_{max}/C_{min}). Nevertheless, positive feedback is always applied with some caution [WH95], [SPJT98], [LG98], [PP994].

Feedback has been exploited in the design of amplifiers since the early days of vacuum-tube electronics. Feedback theory was developed by electronic engineers to satisfy the demand for amplifiers and repeaters exhibiting stable performance for telephone applications. In particular, Black, who was an electronic engineer at Bell Laboratories, is credited as the revolutionary inventor of the feedback amplifier in 1927 [B34]. Since then,

---

[1] The definition of stability and related issues are presented in the next chapter.

feedback has become a key design issue both in analog and digital electronic circuits and systems.

Feedback applied around an analog network allows gain to be stabilised (desensitised) with respect to variations in circuit elements, and active device model parameters. This desensitisation property is crucial in view of parametric uncertainties caused by device parameters spreads, power supply variations, temperature changes, aging phenomena, and so on. Feedback allows the input and output resistances of a given circuit to be suitably modified in any fashion desired. It improves the linearity of the output signal by reducing dependence on the parameters of inherently nonlinear active devices used to implement the open loop circuit. Finally, it can lead to an increase in the closed-loop bandwidth.

However, all these features are paid for in terms of a proportional reduction in gain. This is usually a small price to pay, particularly in applications using operational amplifiers whose dc open-loop voltage gain is very large (60 dB, at least). Besides, as already mentioned, (negative) feedback can determine oscillation, hence, frequency compensation is usually mandatory[2].

3.1 METHOD OF ANALYSIS OF FEEDBACK CIRCUITS

There are several approaches for analysing feedback circuits [H92], [PC981]. The most straightforward one is to directly analyse the circuit by writing the Kirchhoff equations on the small-signal circuit and deriving the circuit characteristics. However, this approach is computationally tedious, it does not allow one to disclose the general properties of feedback circuits which can greatly simplify the analysis and, perhaps more importantly, it gives no insight into circuit behavior and hardly provides useful design equations. Thus, alternative approaches have been developed.

Most of the traditional techniques used to analyse feedback amplifiers start from the idealised block diagram in Fig. 3.1 [G85], [SS91], [C91], [LS94], where blocks **A** and **f** are the open-loop network (that we may identify as the open-loop amplifier) and the feedback network, respectively. Then parameter A represents the gain of the open-loop amplifier and f is the feedback factor, i.e. the portion of the output signal fed back to the input. Signals x_s, x_o, x_f, and x_i represent source, output, feedback and error signals, respectively. This representation gives only approximate results when applied to real amplifiers, principally because it assumes unidirectional blocks. Moreover, it is difficult to take into account the loading effects of the

---

[2] Frequency compensation is treated in Chapter 5 of this book.

feedback network on the basic amplifier, and the determination of A and f is usually a non-trivial task (especially because A is rarely independent of f in practical electronics). Nevertheless, this scheme is suitable for a simple focus on the main feedback properties.

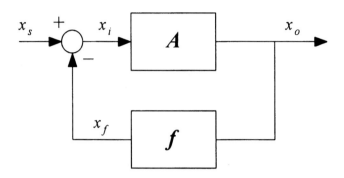

Fig. 3.1. Idealised feedback system.

Assuming blocks **A** and **f** are linear and unidirectional, the transfer function of the feedback amplifier is given by

$$G_F = \frac{x_o}{x_s} = \frac{A}{1 + fA} \tag{3.1a}$$

where G_F is the closed-loop gain and quantity fA is called the *loop gain*. Note that for negative feedback, the loop gain must be positive. Therefore, the closed-loop gain is smaller than the open-loop gain (by a factor $1 + fA$). If $fA < 0$ the feedback is positive. If fA is much greater than 1, the closed-loop transfer function falls to $1/f$, which is independent of the open-loop gain. To better represent this concept we can rewrite (3.1a) as

$$G_F = \frac{1}{f} \frac{fA}{1 + fA} \tag{3.1b}$$

If the feedback network contains circuit elements that are not susceptible to uncertainties we can achieve a closed-loop gain which is desensitised to parameter variations of the open-loop function. This property can be formally expressed by evaluating the sensitivity of the closed-loop gain versus the open-loop one

$$S_A^{G_F} = \frac{dG_F}{dA} \frac{A}{G_F} = \frac{1}{1 + fA} \tag{3.2}$$

which shows that a relative variation in the open-loop system **A** corresponds to a relative variation in the closed-loop system $(1+fA)$ times lower.

In real cases, unfortunately, blocks **A** and **f** are made up of active and passive components, and generally they cannot be assumed to be unidirectional. Of course, to take into account the electrical nature of real amplifiers a straightforward nodal analysis can be applied, but the approach is tedious because it requires the simultaneous solution of numerous equations even for medium-complexity circuits. A commonly used approach models the amplifier and the feedback network with their proper two-port models, each having as their input and output variables both the port voltage and port current [G85]. This leads to an explosion of cases that depend on the specific kind of amplifier and feedback network implemented. The approach can also take into account the non-unidirectional nature of real blocks, but in this case the two-port model increases in complexity and the final analysis, though accurate, becomes extremely elaborate [GM93]. As a result, there are several limitations to this approach. The two-port network which models the block must be selected judiciously. The computation of closed-loop parameters (transfer function and input and output resistances) is tedious, especially if block **A** is a multistage amplifier or a multi-loop feedback is implemented. Moreover, the method is straightforwardly applicable to only those circuits that implement a *global feedback* (a feedback between the input and the output) [PG981], whereas many feedback amplifiers exploit only *local feedback*[3].

Other methods to analyse feedback amplifiers are based on Mason's signal flow graph (SFG) theory [M53], [M56], [MZ60], [C91], [MG91]: they can either be derived from, or related to it [R74], [C90], [B91], [MG91]. The implicit drawback of the uncritical application of the classical signal flow analysis is that we almost completely lose our understanding of circuit behavior, and as a consequence, we have greater difficulties in carrying out the design. This drawback is overcome by approaches that can be derived from signal flow analysis, such as the Rosenstark method [R74], the Choma method [C90], and the Blackman theorem [B43]. The Rosenstark and the Choma methods primarily focus on the evaluation of the closed-loop transfer function. The Blackman theorem –intrinsic to the other two methods– involves the computation of the input and output resistance of a feedback amplifier. Both the Rosenstark and Choma procedures together with the Blackman theorem give circuit designer a very powerful tool for the analysis and design of feedback amplifiers. Indeed, not only do they achieve *exact* relationships by describing the closed-loop amplifier efficiently and in

---

[3] Local feedback occurs when the input and output terminals of the feedback network do not coincide respectively with the output and input terminals of the amplifier.

a very simple manner, but also provide useful design guidelines by highlighting the properties and limitations implicit in specific types of feedback configurations.

3.2 SIGNAL FLOW GRAPH ANALYSIS

Classical SFG techniques have never gained popularity not only in circuit design but even for analysis. This is mainly due to uncertainties in transcribing circuit diagrams into their SFG equivalents. Mason himself recognized that the construction of a SFG is somewhat obscure, and that "A link in the chain of dependency is limited in extent only by one's perception of the problem." Although some of these drawbacks have been overcome [K00], the method still remains less direct than those proposed by Rosenstark and Choma, which will be the only one comprehensively considered in this text and described in the following sections. However, since these methods descend from SFG through which they can easily be demonstrated, we need to introduce some elementary concepts of SFG analysis.

A signal flow graph allows us to graphically represent a circuit (or more generally, a system) through the links between system variables. The nodes on the graph represent variables and the relations among them consist in the branches between the nodes with their associated weights. As a result, a variable is the linear superposition of the node variables at the source of incoming branches.

A general linear circuit can be represented by the signal flow graph shown in Fig. 3.2. Variables x_s and x_o represent the input and output signals, moreover, two other generic variables, x_i and x'_i, linked together through the control (or _critical_) parameter P, are explicitly shown. Parameters a_{ii} are the weight branches. Variables x'_i, x_i and the control parameter, P, can model a controlled generator, or the relation between voltage and current across two nodes of the circuit. This representation is particularly suited to feedback circuits.

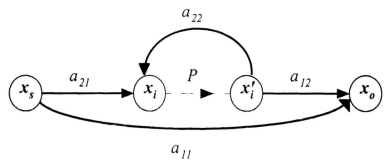

Fig. 3.2. Example of a signal flow graph.

Assuming x_o and x_i as dependent or output variables we can express them as a function of the two other independent or source variables, x_s and x'_i

$$x_o = a_{11}x_S + a_{12}x'_i \qquad\qquad (3.3a)$$

$$x_i = a_{21}x_S + a_{22}x'_i \qquad\qquad (3.3b)$$

Since $x'_i = Px_i$, solving for x_i yields

$$x_i = \frac{a_{21}}{1 - Pa_{22}}x_s \qquad\qquad (3.4)$$

and by substituting (3.4) into (3.3a), the transfer function of the closed-loop amplifier results

$$G_F = \frac{x_o}{x_s} = a_{11} + \frac{a_{12}a_{21}P}{1 - Pa_{22}} \qquad\qquad (3.5)$$

Of course, to evaluate G_F we have to relate weights a_{ii} with circuit elements. Term a_{11} can be found by setting the control parameter to zero and evaluating the transfer function between input and output under this condition. Term a_{12} is the transfer function between the output and the controlled variable, x'_i, setting the input source, x_s, to zero. Term a_{21} can be computed via the transfer function between the source variable and the inner variable, x_i, when the controlled variable, x'_i, is set to zero, which in other words means setting control parameter, P, equal to zero. Term a_{22} gives the relation between the independent and the controlled inner variables setting control parameter, P, and input variable, x_s, to zero. Finally, it is worth noting that, as apparent from Fig. 3.2, product $-Pa_{22}$ represents the loop gain of the network, also more properly termed the *return ratio*.

Equation (3.5) is only one of the many mathematical representations of a linear circuit, which also depends on the particular choice of parameter P. Note, however, that unless P is selected as feedback factor f, which is not always transparent in feedback architectures, expressions for the loop gain and the open loop gain of the feedback amplifiers remain obscure. In the following we utilise (3.5) as a starting point to derive the Rosenstark, Choma, and Blackman procedures.

3.3 THE ROSENSTARK METHOD

Now we will try to express the transfer function G_F, obtained by considering the signal flow analysis, in a form which is more similar to the result given in (3.1b). Indeed, (3.5) can be rewritten as

$$G_F = \frac{a_{11} + \left(\dfrac{a_{12}a_{21}}{a_{22}} - a_{11} \right) P a_{22}}{1 - P a_{22}} \tag{3.6}$$

This form is exactly that found by Rosenstark and allows us to elaborate its procedure. Except for term a_{11} in the numerator, it is similar to relationship (3.1b).

The Rosenstark method is based on the calculation of the _return ratio_, T, the _asymptotic term_, G_∞, and the _direct transmission term_, G_o. All these quantities, must be calculated with respect to one and only one controlled source within the feedback amplifier. Moreover, they are a function of the specific input and output conditions, which means they depend on the input source resistance, R_S, and output load resistance, R_L. The exact input-output transfer function of the feedback amplifier is thus given by [R74]

$$G_F = \frac{G_o + G_\infty T}{1 + T} \tag{3.7}$$

The three quantities in (3.7) are directly related to the weights a_{ii} and parameter P of the flow graph through the following relationships obtained by comparing (3.6) and (3.7)

$$G_o = a_{11} \tag{3.8a}$$

$$T = -P a_{22} \tag{3.8b}$$

$$G_\infty = -\frac{a_{12}a_{21}}{a_{22}} + a_{11} \tag{3.8c}$$

Hence, to evaluate the three terms, we have to refer a controlled source quantity, x'_i, to a controlling quantity, x_i, by a parameter P and follow the steps given below:

1.   Switch off the critical controlled source setting $P = 0$.To achieve the *direct transmission term*, G_o, compute the transfer function between the input and output (i.e., evaluate a_{11}). Condition $P = 0$ requires an open circuit (short circuit) to replace the branch containing the controlled source if P is associated with a controlled current (voltage) source.

2.   Set the input source to zero. This means short-circuiting the voltage source or opening the current source. Replace the critical controlled voltage (current) source by an independent voltage (current) generator of value P. The *return ratio*, T, coincides with the resulting controlling quantity changed in sign, $-x_i$ (i.e., evaluate a_{22} and multiply it to P).

3.   Set the critical parameter to infinitely large (i.e., $P \to \infty$). Since the controlled variable must be finite, this is equal to setting $x_i = 0$. The *asymptotic gain*, G_∞, is the transfer function between the input and the output under this special condition.

A comparison of (3.1) with (3.7) suggests that, for those cases where the direct transmission term, G_o, is negligible, the return ratio equals the product between the amplifier gain and the feedback factor (i.e., $T = fA$). For this reason we will use the terms return ratio and loop gain almost interchangeably, although this is not exactly true. The term $G_o/(1+T)$ in (3.7) can be viewed as a corrective term, which modifies (3.1) when the loop gain is not very large compared to unity. Under this condition, (3.1) and its consequences are no longer valid. Thus (3.7) is a more general, and insightful relation, for computing the closed-loop gain than (3.1). The asymptotic gain equals the reciprocal of the feedback factor (i.e., $G_\infty = 1/f$). Term G_∞ represents the transfer function of a feedback amplifier under the ideal condition of infinite loop gain. Thus, for well-designed feedback circuits, exhibiting low values of G_o and high values of T, the transfer function of the feedback circuit is well approximated by G_∞. The reader can recognise in this observation the basis for the customary paper-and-pencil analysis of feedback configurations employing ideal operational amplifiers.

In order to illustrate the use of the Rosenstark theory and to give an idea of its strength and simplicity, let us apply the method to a common Z configuration whose load determines an intrinsic feedback. The circuit, reported in Fig. 3.3a, is the same as that analysed in Chapter 2 (the source and the load resistances are coincident with resistance R_Y and $R_X//R_L$, respectively). The small signal model, in which for simplicity the bulk transconductance, g_{mb}, and the transistor output resistance, r_o have been neglected, is shown in Fig. 3.3b.

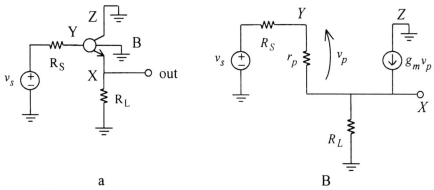

Fig. 3.3. Common Z configuration: a) AC schematic, b) small-signal model.

Assume as control parameter P the transconductance of the transistor, g_m. Hence, the controlling variable (denoted before as x_i) is voltage v_p, and the controlled variable (x'_i) is the current of the controlled generator. Setting parameter P equal to zero means switching off the controlled current source. By evaluating under this condition the transfer gain of the amplifier we found the direct gain G_o. In particular, the voltage at node X is due to a partition of the input signal,

$$G_o = \frac{R_L}{R_L + r_p + R_S} \tag{3.9}$$

Consider the controlled current source an independent current source, $P = g_m$, and set to zero (i.e. shorting) the input signal generator. Now evaluate the voltage v_p due to the independent current source, P,

$$v_p = -r_p \frac{R_L}{R_L + r_p + R_s} g_m \tag{3.10}$$

Hence, changing the sign and multiplying for the transconductance value we obtain the return ratio

$$T = g_m r_p \frac{R_L}{R_L + r_p + R_S} \tag{3.11}$$

Finally, set control parameter, P, equal to infinity. The controlled current source is finite, and this means that the control variable, v_p, must be equal to zero. Hence, $i_p = i_s = 0$, and the asymptotic gain results

$$G_\infty = 1 \tag{3.12}$$

Substituting (3.9), (3.11) and (3.12) in (3.7), we get

$$A_v = \frac{\left(1 + g_m r_p\right) R_L}{\left(1 + g_m r_p\right) R_L + r_p + R_S} \tag{3.13}$$

which gives exactly the same result found directly in paragraph 2.6. To immediately verify this assertion set $R_S = 0$ in (3.13) and compare it with (2.28a).

Besides the simplicity of the procedure, the Rosenstark method, gives precise information on the ideal behavior of the common Z amplifier through the asymptotic gain and illuminates how an increase in the loop-gain, T, moves the final transfer function close to its ideal value G_∞.

3.4 THE CHOMA METHOD

Like the Rosenstark method, the one proposed by Choma starts from the signal flow representation expressed by relationship (3.6), but adopts as reference the relation in (3.1a), according to

$$G_F = a_{11} \frac{1 - P\left(a_{22} - \dfrac{a_{12} a_{21}}{a_{11}}\right)}{1 - P a_{22}} \tag{3.14}$$

In this case, only the denominator is expressed in terms of loop gain, T. The whole function is multiplied for weight a_{11}, which represents the Direct Transmission Term, G_o, and the numerator now depends on a novel parameter termed the *null return ratio*, T_R. Thus the exact input-output transfer function of the feedback amplifier is given by [C91]

$$G_F = G_o \frac{1 + T_R}{1 + T} \tag{3.15}$$

where, of course, (3.8a) and (3.8b) hold, and

$$T_R = -P\left(a_{22} - \frac{a_{12} a_{21}}{a_{11}}\right) \tag{3.16}$$

Like for the Rosenstark method we have to choose a controlled source P inside the feedback circuit. To calculate the Return Ratio, T, and the Direct Transmission Term, G_o, we can follow the same steps described in point 1) and 2). But now, we have to change step 3) to evaluate the null return ratio. The new point 3) is

3. Replace the critical controlled source by an independent source of value P (like in the second point of the previous paragraph), without nullifying the input source. The Null Return Ratio, T_R, will be coincident with the resulting controlling quantity changed in sign, $-x_i$, assuming the output voltage is equal to zero. It is worth noting that the input source is not independent, but its value must guarantee the zero condition at the output.

To demonstrate point 3), set $x_o = 0$ in eq. (3.3a), yielding

$$x_s = -\frac{a_{12}}{a_{11}} x'_i \tag{3.17}$$

and after substituting eq. (3.17) in (3.4) we get

$$x_i = \left(a_{22} - \frac{a_{12}a_{21}}{a_{11}} \right) x'_i \tag{3.18}$$

Like the asymptotic gain, also the null return ratio gives interesting information from a circuit/design point of view. Moreover, it helps to identify the nature of the feedback. The ratio between the return ratio and the null return ratio, T/T_R, quantifies the degree to which the local feedback approaches global feedback [C91]. When it is ∞ the feedback is global.

Of course, both Rosenstark and Choma methods give the same results, and comparing (3.7) with (3.15) we obtain

$$\frac{T}{T_R} = \frac{G_o}{G_\infty} \tag{3.19}$$

To evaluate the null return ratio, consider again the small-signal model of the common Z amplifier in Fig. 3.3b. Since voltage at node X must be assumed to be zero, this means that the current of the critical generator P (equal to g_m) all flows through resistances R_S and r_p. Hence the null return ratio is given by

$$T_R = -v_p = g_m r_p \tag{3.20}$$

3.5 THE BLACKMAN THEOREM

Feedback not only modifies the closed-loop network transfer function, but also determines a change in the input/output resistances (impedances). Specifically, as demonstrated in many classical textbooks such as [MG91], the input resistance increases (decreases) by a factor $1+T$ when the signal mixed at the input is a voltage (current). In contrast, the output resistance increases (decreases) by a factor $1+T$ when the sensed signal is a current (voltage). This conforms to the well-known rule that a series (shunt) feedback connection in either the input or output increases (decreases) the associated port resistance. However, in practical circuits there is not often a sharp separation between the above mentioned cases. In other words, there are situations that do not match either of the canonical configurations, and hence a more general technique is needed to derive resistance relations.

Input and output resistances can be efficiently evaluated by using the Blackman theorem [B43]. It was introduced in 1943 and rediscovered by Rosenstark [74] and Grabel [MG91], and was also independently developed through signal flow analysis by Choma [C90].

The signal flow scheme in Fig. 3.2 and related equations (3.3) and (3.4) are general and, hence, can be used to represent the relation between the voltage and the current at the input or the output port. Let variables x_o and x_s be the voltage and the current of the considered port respectively. Eqs. (3.3) and (3.4) can be rewritten as

$$V = a_{11}I + a_{12}x'_i \tag{3.21a}$$

$$x_i = a_{21}I + a_{22}x'_i \tag{3.21b}$$

Following the same procedure which leads to the Choma representation we get

$$R = \frac{V}{I} = a_{11}\frac{1 - P\left(a_{22} - \dfrac{a_{12}a_{21}}{a_{11}}\right)}{1 - Pa_{22}} \tag{3.22}$$

To interpret this result observe that term a_{11} is the ratio between V and I, when the controlled variable is zero. In other words, this means that term a_{11} is the resistance level without feedback (i.e., $P = 0$). Term Pa_{22} is the loop gain, which is computed by setting variable x_s and, hence, I equal to zero. This means evaluating the loop gain with the considered port unloaded. Finally, as derived in the Choma approach, term $P(a_{22} - a_{12}a_{21}/a_{11})$ is the loop gain setting the output variable to zero. This means that the null return ratio is the loop-gain after short-circuiting the port considered.

From the above discussion, and representing the loop gain as an explicit function of R_S and R_L, we can specify (3.22) to represent the closed-loop input and output resistances, r_{in} and r_{out}, as follows

$$r_{in} = r_{in,ol} \frac{1+T(0,R_L)}{1+T(\infty,R_L)} \tag{3.23}$$

$$r_{out} = r_{out,ol} \frac{1+T(R_S,0)}{1+T(R_S,\infty)} \tag{3.24}$$

where $r_{in,ol}$, and $r_{out,ol}$, are the corresponding driving point input and output resistances with the critical parameter P equal to zero, and $T(0, R_L)$, $T(\infty,R_L)$, $T(R_S, 0)$ and $T(R_S, \infty)$ are the return ratios under the specific conditions for the source resistance, R_S, and load resistance, R_L.

We have already discussed the ability of feedback to provide gain desensitisation. However, such desensitisation of the input and output resistances cannot be achieved. In fact, it can be shown through relations (3.23) and (3.24) that I/O resistances depend on open loop parameters.

The application of Blackman theorem to the common Z amplifier in Fig. 3.3 is now described. Assume transconductance g_m as the control parameter P. The input and output[4] resistances with $g_m = 0$ are

$$r_{in,ol} = r_p + R_L \tag{3.25}$$

$$r_{out,ol} = r_p + R_S \tag{3.26}$$

Both $T(\infty, R_L)$ and $T(R_S, 0)$ are equal to zero, whereas $T(0, R_L)$ and $T(R_S, \infty)$ result

$$T(0,R_L) = \frac{g_m r_p R_L}{r_p + R_L} \tag{3.27}$$

$$T(R_S,\infty) = g_m r_p \tag{3.28}$$

Hence, input and output resistances are given by

---

[4] As usual, the output resistance is calculated excluding resistor R_L.

$$r_{in} = r_p + R_L \left(1 + g_m r_p \right) \tag{3.29}$$

$$r_{out} = \frac{r_p + R_S}{1 + g_m r_p} \tag{3.30}$$

Of course the two results found here are the same as those obtained by simplifying (2.24) and (2.25).

Chapter 4

STABILITY
FREQUENCY AND STEP RESPONSE

In the previous chapter we enunciated the basic feedback concepts and described efficient techniques for analysing feedback amplifiers. Particularly, we defined the open-loop gain, the loop-gain (or return ratio), and other quantities, all as *DC values*. However, these quantities are in general a function of frequency and they should be better referred to as *transfer functions* instead of gains. Moreover, the feedback factor could also be frequency dependent (to this end, the best example is perhaps the well-known RC-active integrator made up of an op-amp and a feedback network constituted by a resistor and a capacitor). Thus, all these effects should be taken into account in the Rosenstark and Choma relationships, (3.8) and (3.16), which allow us to accurately obtain the closed-loop transfer function. In addition, for a first-order model, they should also be considered in (3.1). Similarly, in the Blackman equations, (3.23) and (3.24), we need to use the appropriate return ratio transfer function to obtain input and output impedances instead of resistances.

For the sake of simplicity, in this chapter we will assume that the feedback factor is constant, at least in the frequency range of interest. In addition, we will assume that the feedback network is designed so as to not introduce further poles in the loop gain. Such a condition is fortunately often verified in feedback amplifiers with a purely resistive feedback network.

It should be well known to the reader that an electronic circuit and system are said to be *stable* if all bounded excitations yield bounded responses. Otherwise, if bounded excitations produce an unbounded response the system is said to be *unstable*. Passive RLC circuits are by nature stable. Active networks contain internal energy sources that can combine with the input excitation to cause the output to increase indefinitely or sustain oscillations. Note, however, that in practice the output of an unstable circuit cannot diverge *indefinitely*, since a limit is set by the power supply rails.

It should also be well known that stability is ensured if all the poles of a given circuit/system lie in the left-half of the *s*-plane. Thus, we could check the stability of a feedback amplifier by evaluating the closed-loop transfer function and determining the locations of its poles. This procedure, however, does not provide design insides and does not specify the margins by which stability is achieved. In fact, circuit components are affected by manufacturing tolerances, temperature and ageing phenomena, etc., which cause a parameter to deviate from its nominal value. Under this scenario, we need to introduce safety stability margins, which are the *phase margin* and *gain margin*. Moreover, even stable amplifiers, hence that have a bounded response, can take too much time to reach a steady state. For this purpose, the classical feedback circuit analysis technique derived from the well-known Bode disclosures can be utilised [B45].

In the following paragraphs we will examine the frequency response of transfer functions characterised by different combinations of poles (and zeros) that are found usually in real practice. Starting from this, useful definitions will be given which help designers to derive fundamental relations to ensure closed-loop stability with adequate margins. The closed-loop step response in the time domain, for each typology of transfer function, is also derived.

4.1 ONE-POLE FEEDBACK AMPLIFIERS

Among the feedback properties, the closed-loop bandwidth extension to the original open-loop amplifier is often included [G85], [SS91]. We will show that this property applies only to one-pole amplifiers, but is not effective in multi-pole amplifiers.

Let us consider an open-loop amplifier having the following transfer function including a single (negative) pole, whose angular frequency is p_1

$$A(s) = \frac{A_o}{1 + \dfrac{s}{p_1}} \tag{4.1}$$

Now connect the amplifier in feedback with a pure resistive network, whose feedback factor is f, as shown in Fig. 4.1.

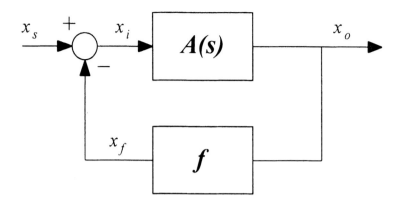

Fig. 4.1. Block diagram of a feedback amplifier.

Returning to (3.1a), the closed-loop transfer function results as

$$G_F(s) = \frac{G_{Fo}}{1 + \dfrac{s}{p_{F1}}}$$ (4.2)

where G_{Fo} is the DC closed-loop gain and p_{F1} is the closed-loop pole, each given by

$$G_{Fo} = \frac{A_o}{1 + fA_o} = \frac{1}{f}\frac{T_o}{1 + T_o} \approx \frac{1}{f}$$ (4.3a)

$$p_{F1} = (1 + fA_o)p_1 = (1 + T_o)p_1 \approx T_o p_1$$ (4.3b)

The foregoing approximations hold for large loop gains ($T_o \gg 1$), which are required for an adequate desensitisation of the closed-loop response with respect to open-loop parameters. It is seen that increasing T_o from zero shifts the pole along the negative real axis, as illustrated in Fig. 4.2[1]. Since the pole is located in the negative s-plane for any value of f, the system is termed *absolutely* or *unconditionally* stable. This denotes an attractive condition indicating that a one-pole amplifier is stable under all input signal conditions

---

[1] This plot is called the root locus diagram. Its construction can become tedious for higher order systems and we do not make use of this tool to examine stability. The interested reader is referred to [SS91], [G85], or any feedback control text e.g. [FPE94].

and for all ranges of component values. Unfortunately, this is not a realistic case, since real amplifiers have more than one pole.

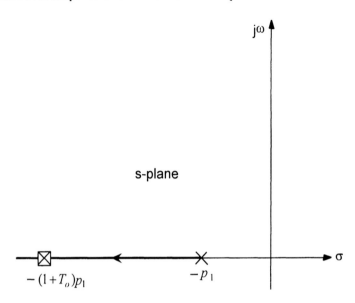

Fig. 4.2. Effect of feedback on the pole location for a single-pole amplifier.

Returning to (4.3b), we see that the closed-loop pole has been shifted to a higher frequency by a factor equal to $1+T_o$ (that is, approximately the DC loop gain), which is the same amount of reduction experienced by the closed-loop DC gain with respect to the open-loop gain. Thus, a gain bandwidth trade-off exists between the open- and closed-loop transfer functions, indicating that in a one-pole amplifier we can apply feedback to obtain higher bandwidth where amplifier gain reduction is allowed. This trade-off is represented by the *gain-bandwidth product, ω_{GBW}*, which is the product of the DC open-loop gain, A_o, and its –3-dB angular frequency p_1. Note also that the gain-bandwidth product of a single-pole function exactly equals its *unity-gain frequency, ω_T* (i.e., the frequency at which the module of the gain becomes unitary, for this reason it is also called the *transition* frequency). Moreover, ω_{GBW} is an invariant amplifier parameter, since its value is the same for the open-loop and closed-loop amplifier, as illustrated in Fig. 4.3, showing the open-loop, closed-loop and loop-gain transfer functions. Of course, the gain-bandwidth product of $A(s)$ is independent of the degree of feedback applied and is equal to the maximum bandwidth achieved with the unitary feedback factor, $f = 1$ (i.e., with the amplifier in unity gain feedback configuration). More interestingly, (4.3b) predicts that the gain-bandwidth product of the loop-gain transfer function will equal the closed-loop pole. Thus, when studying the stability of a feedback amplifier, it is usually convenient to analyse the frequency response of the loop-gain rather than that of the open-loop transfer function. This is because the loop-gain transfer function gives information on most of the closed-loop properties.

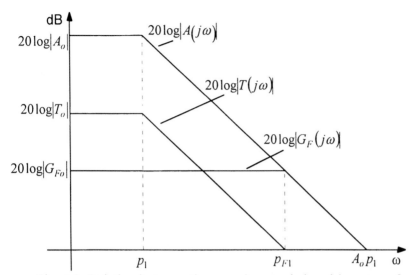

Fig. 4.3. Relation between the open-loop and closed-loop transfer function magnitudes.

The closed-loop characteristics of an amplifier can be also investigated in the time domain, by evaluating the response, $x_o(t)$, to a unitary input step $u(t)$. The step response gives important specifications for applications (such as instrumentation, control, and sample data systems) sensitive to the amplifier's transient response.

Let us consider a closed-loop configuration comprising the single-pole amplifier with the transfer function shown in (4.1). The step response is easily found to be

$$x_o(t) = \left(1 - e^{-p_{F1}t}\right) G_{Fo} u(t) \qquad (4.4)$$

The output steady state value is G_{Fo} and is reached exponentially with time constant $1/p_{F1}$.

The reader should know that the _settling time_, t_s, is the time interval required for the output response to settle to some specified percentage of the final value. For a single-pole amplifier the settling time is then proportional to p_{F1}. Usually, the settling time needed to reach 1% or 0.1% of the final value is considered. In these two cases, the settling time results $4.6/p_{F1}$ and $6.9/p_{F1}$.

4.2 TWO-POLE FEEDBACK AMPLIFIERS

The loop-gain transfer function of real amplifiers includes more than one single pole. In the absence of suitable compensation, this can cause instability phenomena even under negative feedback. To demonstrate these instability problems, and the related importance of a sufficiently large separation between the two lowest poles of the loop-gain transfer function, consider now an open-loop amplifier with two real negative poles. As already mentioned, and as will be further explained in the following chapters, it is more convenient to analyse the loop gain instead of the amplifier open-loop gain.

Assume that the amplifier operating within the given feedback network gives rise to the following two-pole loop-gain function

$$T(s) = \frac{T_o}{\left(1 + \dfrac{s}{p_1}\right)\left(1 + \dfrac{s}{p_2}\right)} \tag{4.5}$$

It should now be observed that for second-order and higher-order transfer functions, the gain-bandwidth product, $\omega_{GBW} = T_o p_1$, does not necessarily coincide with the unity-gain frequency ω_T. However, this is still a good approximation if the second pole is greater than ω_{GBW}. This observation is graphically explained in Fig. 4.4, where two loop-gain functions (with different pole separation) are plotted.

Note that hereinafter and unless differently indicated, we will denote the gain-bandwidth product of the loop-gain transfer function with ω_{GBW}

Using (3.1a), the closed-loop transfer function becomes

$$G_F(s) = \frac{G_{Fo}}{1 + 2\dfrac{\xi}{\omega_o} s + \dfrac{s^2}{\omega_o^2}} = \frac{G_{Fo}}{\left(1 + \dfrac{s}{p_{F1}}\right)\left(1 + \dfrac{s}{p_{F2}}\right)} \tag{4.6}$$

where ω_0 and ξ are the parameters called *pole frequency* and *damping factor*, respectively, and are expressed as functions of the loop-gain poles by [SS91]

$$\omega_o = \sqrt{p_1 p_2 (1 + T_o)} \tag{4.7}$$

$$\xi = \frac{p_1 + p_2}{2\omega_o} \approx \frac{1}{2\sqrt{T_o}}\left(\sqrt{\frac{p_1}{p_2}} + \sqrt{\frac{p_2}{p_1}}\right) \tag{4.8}$$

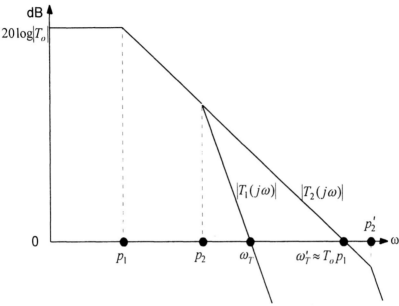

Fig. 4.4. Magnitude of two second-order loop gain transfer functions characterised by two different second poles p_2 and p_2'. Only in the second case does the gain-bandwidth product coincide with the transition frequency.

The last identity in (4.6) is an alternative expression of $G_F(s)$, with p_{F1} and p_{F2} as the closed-loop amplifier poles given by

$$p_{F1,2} = \omega_o\left(\xi \pm \sqrt{\xi^2 - 1}\right) \tag{4.9}$$

These poles are either real or complex conjugate pairs, depending on the value of ξ (or parameter Q equal to $1/(2\xi)$, sometimes used instead of ξ and called the _pole Q factor_). The location of the closed-loop poles, as the DC loop-gain is increased from zero, is illustrated in Fig. 4.5 showing that a second-order feedback system is absolutely stable. However, the design of a second-order system having a specific and well-defined frequency and transient response requires careful consideration of where the poles are to be located.

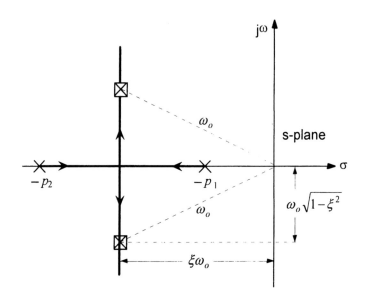

Fig. 4.5. Effect of feedback on the pole location for an amplifier with two real poles.

By normalising the module of the closed-loop transfer function and the angular frequency to G_{Fo} and ω_o, respectively, we obtain the frequency responses plotted in Fig. 4.6.

This figure helps to visualise that when ξ is lower than a critical value (namely $1/\sqrt{2}$), an overshoot in the frequency domain arises at a frequency, ω_{cp}, with the (peak) amplitude, G_{Fp}, both given below [MG91]

$$\omega_{cp} = \omega_o \sqrt{1 - 2\xi^2} \tag{4.10}$$

$$G_{Fp} = G_{Fo} \frac{1}{2\xi\sqrt{1-\xi^2}} \tag{4.11}$$

It is apparent that the relative amplitude of the overshoot depends only on the damping factor, ξ. For $\xi = 1/\sqrt{2}$ it can be shown that the module of the frequency response is maximally flat (which is often referred to as the *Butterworth* condition). Specifically, this condition yields the largest possible closed-loop 3-dB bandwidth within the constraint of a monotone decreasing frequency response.

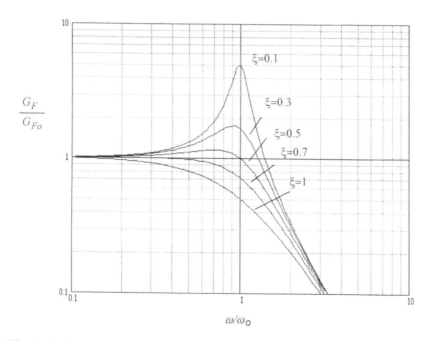

Fig. 4.6. Normalised module of the closed-loop frequency-response of a two-pole amplifier, for different values of the damping factor.

The expression of the step response of the closed-loop two-pole amplifier is

$$x_o(t) = \left[1 - \frac{1}{p_{F1} - p_{F2}}\left(p_{F2}e^{-p_{F1}t} - p_{F1}e^{-p_{F2}t}\right)\right]G_{Fo}u(t) \qquad (4.12)$$

If the closed-loop poles are complex conjugate –a condition which arises when the value of ξ is lower than 1– the step response exhibits an *underdamped* behavior (conversely, an *overdamped* closed-loop response requires $\xi > 1$). In such cases the step response is better expressed by the following relationship

$$x_o(t) = \left\{1 - \left[\frac{\xi}{\sqrt{1-\xi^2}}\sin\left(\sqrt{1-\xi^2}\,\omega_o t\right) + \cos\left(\sqrt{1-\xi^2}\,\omega_o t\right)\right]e^{-\xi\omega_o t}\right\}G_{oF}u(t) =$$

$$= \left\{1 - \left[\sqrt{\frac{1}{1-\xi^2}}\cos\left(\sqrt{1-\xi^2}\,\omega_o t - \frac{\xi}{\sqrt{1-\xi^2}}\right)\right]e^{-\xi\omega_o t}\right\}G_{oF}u(t) \qquad (4.13)$$

Underdamped amplifiers are not unstable systems, but nonetheless they are usually unacceptable, because overshoot arises in the time domain which is responsible for slow settling behaviour.

Normalising the step response to $u(t)$, we can draw the plots in Fig. 4.7, illustrating the step response of a two-pole feedback amplifier for different values of ξ, versus $\omega_o t$.

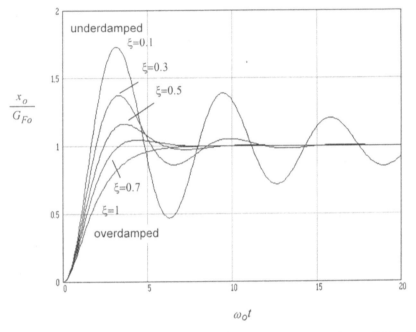

Fig. 4.7. Normalised step response of a two-pole feedback amplifier for different values of the damping factor.

To maintain peaking in both the frequency and step responses below a desired value, parameter ξ must be properly set. To this end, relationship (4.8) implicitly provides the required relation between the two (open-loop) poles for a given value of ξ and T_o. In order to avoid excessive underdamping, open-loop amplifiers must be designed with a dominant pole and a second pole at a frequency higher than the gain-bandwidth product of the loop gain (i.e., $p_2 > T_o p_1$). Thus, to analyse and design feedback amplifiers, it is useful to introduce a new parameter called the *separation factor*, K, which is the ratio between the second pole and the gain-bandwidth product of the return ratio $T(s)$

$$K = \frac{p_2}{\omega_{GBW}} = \frac{p_2}{T_o p_1} \tag{4.14}$$

The separation factor is strictly related to a parameter of the loop gain commonly used to measure the degree of stability of a feedback system namely, the *phase margin*[2], ϕ. Indeed, the phase margin is defined as 180° plus the phase of the return ratio evaluated at the transition frequency, ω_T. Figure 4.8 illustrates how the phase margin is determined on the Bode plots[3] of a second-order transfer function.

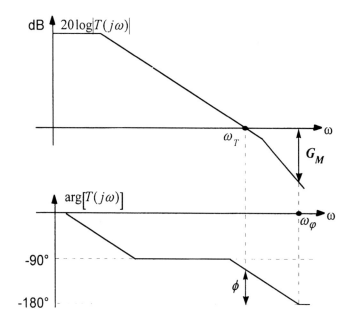

Fig. 4.8. Bode plots of a two-pole system and graphical definition of phase and gain margins.

For a second-order system with negative poles we have

$$\phi = 180° - \tan^{-1}\frac{\omega_T}{p_1} - \tan^{-1}\frac{\omega_T}{p_2} = \tan^{-1}\frac{p_1}{\omega_T} + \tan^{-1}\frac{p_2}{\omega_T} \qquad (4.15)$$

In a well-designed amplifier T_o is larger than unity and the condition $p_2 > T_o p_1$ must also hold. Thus, ω_{GBW} is about equal to the transition frequency, ω_T, and $\tan^{-1}(p_1/\omega_T) \approx 0$. Then (4.14) is reduced to

---

[2] Another parameter, less frequently utilised by electronic designers, is the *gain margin*, G_M, defined as the difference between the gain $20 \log | T(\omega_\varphi)|$ and 0 dB, where ω_φ is the frequency at which the phase equals $-180°$.

[3] We assume that the reader is familiar with the Bode plot technique. For a review of this method see for instance [SS91], [G85].

$$K \approx \tan \phi \tag{4.16}$$

indicating that in a two-pole amplifier K is almost equal to the trigonometric tangent of the phase margin. In other words, for a target phase margin, we obtain through (4.16) the value of the separation factor required during the compensation design step.

From the above it derives that to design and analyse feedback amplifiers it is more convenient to represent the closed-loop transfer function, $G_F(s)$, as a function of the gain-bandwidth product, ω_{GBW}, and the separation factor, K [PP982]. Indeed, the conventional parameters, ω_o and ξ (or parameter Q) traditionally used in feedback systems, have been found very useful in designing and analysing filters, but are less effective in the context of feedback amplifiers. This because, unlike ω_{GBW} and K, which are parameters of the loop gain, ω_o and ξ are parameters related to the closed-loop amplifier. But designer effort is mainly (if not exclusively) focused on properly setting the open-loop amplifier parameters in order to achieve the closed-loop specifications. In addition, the new representation provides a simple vehicle for characterising feedback systems. Indeed, the pole frequency and the damping factor can be expressed as

$$\omega_o = p_1 \sqrt{K T_o \left(1 + T_o\right)} \approx \omega_{GBW} \sqrt{K} \tag{4.17}$$

$$\xi = \frac{1 + K T_o}{2 T_o \sqrt{K}} \approx \frac{\sqrt{K}}{2} \tag{4.18}$$

Upon inserting (4.17) and (4.18) into (4.6), the closed-loop transfer function becomes

$$G_F(s) = \frac{G_{Fo}}{1 + \dfrac{s}{\omega_{GBW}} + \dfrac{s^2}{\omega_{GBW}^2 K}} \tag{4.19a}$$

or equivalently

$$G_F(s') = \frac{G_{Fo}}{1 + s' + \dfrac{s'^2}{K}} \tag{4.19b}$$

where the complex frequency s' is the complex frequency s normalised to ω_{GBW}.

The normalised overshoot frequency and correspondent peak (as functions of ω_{GBW} and K) are now determined to be

$$\omega_{cp} = \omega_{GBW}\sqrt{K(1-K)} \qquad (4.20a)$$

$$\frac{G_{Fp}}{G_{Fo}} = \frac{2}{\sqrt{K(4-K)}} \qquad (4.20b)$$

The magnitude of the frequency response normalised to G_{Fo} versus ω/ω_{GBW} for different values of K, is plotted in Fig. 4.9. It can be noted that condition $K = 2$ i.e., $p_2 = 2\omega_{GBW}$, means a maximally flat frequency response. Moreover, for a given ω_{GBW}, the bandwidth diminishes for decreasing values of K.

The poles of the closed-loop amplifier can be also expressed as

$$p_{F1,2} = \omega_{GBW}\left[\frac{K}{2} \pm \sqrt{\left(\frac{K}{4}-1\right)K}\right] \qquad (4.21)$$

and the response to an input unitary step (assuming underdamped behaviour) is

$$x_o(t) = \left\{1 - \left[\sqrt{\frac{K}{4-K}}\sin\left(\sqrt{K-\frac{K^2}{4}}\omega_{GBW}t\right) + \cos\left(\sqrt{K-\frac{K^2}{4}}\omega_{GBW}t\right)\right]e^{-\frac{K}{2}\omega_{GBW}t}\right\}G_{oF}u(t) =$$

$$= \left\{1 - \left[\sqrt{\frac{4}{4-K}}\cos\left(\sqrt{K-\frac{K^2}{4}}\omega_{GBW}t - \sqrt{\frac{K}{4-K}}\right)\right]e^{-\frac{K}{2}\omega_{GBW}t}\right\}G_{oF}u(t) \qquad (4.22)$$

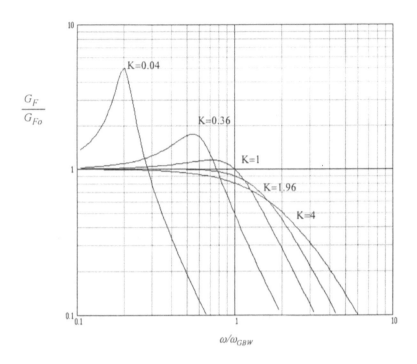

$$\frac{G_F}{G_{Fo}}$$

$$\omega/\omega_{GBW}$$

Fig. 4.9. Normalised module of the closed-loop frequency response for a two-pole feedback amplifier versus normalised frequency, for different values of K.

The step response versus time normalised to ω_{GBW} for different values of K, is plotted in Fig. 4.10.

To optimise the closed-loop amplifier step response, useful information for the designer are the time, t_p, when the *first* peak occurs (i.e., the time at which the first time derivative of $x_o(t)$ becomes zero) and its overshoot, D, [YA90], that are given by

$$t_p = \frac{2\pi}{\omega_{GBW}\sqrt{4K - K^2}} \tag{4.23}$$

$$D = e^{-\pi\sqrt{\frac{K}{4-K}}} \tag{4.24}$$

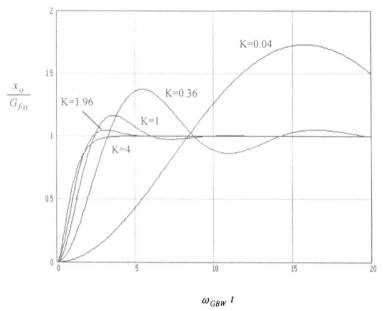

Fig. 4.10. Step response for a two-pole feedback amplifier versus
$\omega_{GBW} t$ for different values of parameter K.

Like for the peaking amplitude in the frequency domain, the overshoot amplitude in the time domain depends only on the value of K. Relationships (4.23) and (4.24) are useful for optimising design in the time domain. Equation (4.24) gives the value of K for a specified settling error, and from (4.23) we determine the gain-bandwidth product needed by the settling time required. For instance, obtaining a step response to within 1% means $K = 2.73$. From (4.16) this value corresponds to a phase margin of about 70°. Then, if 1% settling is to be achieved within a time period not greater than 100 ns, the required gain-bandwidth product is $2\pi \cdot 5.37$ Mrad/s.

It should now be pointed out that in real amplifiers the second pole is generally fixed by design and topology constraints. Subsequently, the requirement on parameter K (or equivalently on the phase margin) indicates the gain-bandwidth we must provide to the loop-gain transfer function to ensure an adequate stability margin. To this end, as shall be discussed in detail in the next chapter, we have to properly reduce the dominant pole of the open-loop amplifier. This mandatory operation drastically reduces the high-frequency capability of the feedback amplifier, which, if operated in open-loop conditions, is characterised by a high-sensitive gain, but has its maximum bandwidth potential limited by the frequency of the second pole. As a consequence, the bandwidth improvement caused by the feedback is effectively achieved only in one-pole amplifiers. However, these are somewhat an abstraction, since real architectures –even single-stage ones– exhibit multiple poles. Bandwidth extension is, therefore, not such a general and effective property as commonly reported. Actually, amplifiers with the highest frequency performance (e.g., RF amplifiers) all adopt open-loop topologies.

4.3 TWO-POLE FEEDBACK AMPLIFIERS WITH A POLE-ZERO DOUBLET

The loop gain of real amplifiers can include a pole-zero doublet beside two significant poles. Usually, a doublet arises from imperfect pole-zero or feed-forward compensation due to process tolerances [KM74], [BAR80], [PP95], or is caused by the frequency limitation of current mirrors when they are used to provide a differential-to-single conversion [GPP99].

The degradation in the settling performance of a one-pole amplifier with a pole-zero doublet was first discussed in [KMG74]. The effect of the doublet in a class AB one-pole amplifier was then analysed for both the settling and slewing time periods in [S91], [SY94]. However, extending the results in [KMG74] to two-pole amplifiers is not as straightforward as sometimes reported [GM74], [LS94], [EH95].

A simpler representation of a two-pole amplifier with a pole-zero doublet was proposed in [PP992]. The approach is based on the consideration that, in practice, the pole and the zero forming the doublet are often very close. In addition, they are usually located at a frequency around or greater than ω_{GBW}. Thus, such a doublet leaves the loop-gain unity-gain frequency almost unchanged, but can considerably alter the phase margin. We now demonstrate that a two-pole amplifier with such a pole-zero doublet can be modeled by an *equivalent* pure two-pole amplifier with a modified second pole.

Consider an amplifier whose loop-gain transfer function includes two poles and a pole-zero doublet (p_d and z_d), as given below

$$T(s) = T_o \frac{1}{\left(1 + \dfrac{s}{p_1}\right)\left(1 + \dfrac{s}{p_2}\right)} \cdot \frac{1 + \dfrac{s}{z_d}}{1 + \dfrac{s}{p_d}} \qquad (4.25)$$

Without loss generality, assume p_1 to be the lowest frequency pole (remember that a dominant-pole behavior is mandatory to achieve stability). Phase margin evaluation of (4.25) gives

$$\phi = \tan^{-1} \frac{p_2}{\omega_T} + \tan^{-1} \frac{p_d}{\omega_T} - \tan^{-1} \frac{z_d}{\omega_T} \tag{4.26}$$

From the considerations regarding the location of p_d and z_d made above, the transition frequency, ω_T, can be assumed to be equal to the gain-bandwidth product, ω_{GBW}, (which formally represents the unity-gain frequency of a one-pole amplifier) so that (4.26) can be rewritten as

$$\phi = \tan^{-1} \frac{p_2}{\omega_{GBW}} + \tan^{-1} \frac{p_d}{\omega_{GBW}} - \tan^{-1} \frac{z_d}{\omega_{GBW}} \tag{4.27}$$

Nevertheless, if we want to accurately evaluate the deviation of ω_T from ω_{GBW} we can use the following results.

By using parameter K defined in (4.14), and the trigonometric identity

$$\tan(a + b + c) = \frac{\tan a + \tan b + \tan c - \tan a \tan b \tan c}{1 - \tan a \tan b - \tan a \tan c - \tan b \tan c} \tag{4.28}$$

from relationship (4.27) we get

$$\tan(\phi) = K \left[1 + \frac{\left(K + \dfrac{1}{K}\right) \dfrac{p_d}{\omega_{GBW}} \Delta}{1 + \left(\dfrac{\omega_d}{\omega_{GBW}}\right)^2 - K \dfrac{p_d}{\omega_{GBW}} \Delta} \right] \tag{4.29}$$

where parameter Δ is the spacing of the doublet normalised to its pole and ω_d is the doublet average frequency (evaluated as the geometric mean between p_d and z_d), respectively, defined by

$$\Delta = \left(1 - \frac{z_d}{p_d}\right) \tag{4.30}$$

$$\omega_d = \sqrt{p_d z_d} \tag{4.31}$$

Normalising the doublet frequency, ω_d to ω_{GBW}

$$\omega_{dn} = \frac{\omega_d}{\omega_{GBW}} \tag{4.32}$$

relationship (4.29) can be written

$$\tan(\phi) = K \left[1 + \frac{\left(K + \frac{1}{K} \right) \omega_{dn} \Delta}{\left(1 + \omega_{dn}^2 \right) \sqrt{1 - \Delta} - K \omega_{dn} \Delta} \right] \tag{4.33}$$

By inspection of (4.33) we note that the second term in square brackets represents the change caused by the doublet in the tangent of the phase margin. Of course, if $\Delta = 0$ pole p_d perfectly matches zero z_d and (4.33) simplifies to (4.16). Moreover if z_d is greater (lower) than p_d, Δ is negative (positive), and the effect of the doublet is to decrease (increase) the phase margin compared to the same two-pole system without the doublet.

From the above it derives that we can model the two-pole amplifier with a pole-zero doublet by using an equivalent pure two-pole amplifier having the same gain-bandwidth product (i.e., gain and dominant pole) and a second pole, p_{2eq}, which guarantees the same phase margin given by (4.33). Hence, (4.25) is approximated by

$$T(s) \approx T_o \frac{1}{\left(1 + \frac{s}{p_1} \right) \left(1 + \frac{s}{p_{2eq}} \right)} \tag{4.34}$$

where

$$p_{2eq} = K \left[1 + \frac{\left(K + \frac{1}{K} \right) \omega_{dn} \Delta}{\left(1 + \omega_{dn}^2 \right) \sqrt{1 - \Delta} - K \omega_{dn} \Delta} \right] \omega_{GBW} \tag{4.35}$$

The second term within square brackets in (4.35) gives the deviation of the equivalent second pole with respect to the actual second pole, which depends on both ω_{dn} and Δ. It can be easily shown that the deviation is at a maximum when $\omega_{dn} = 1$, for a fixed value of Δ (and K), that is when the average doublet frequency is equal to the gain-bandwidth product. In contrast, with ω_{dn} and K given, the influence of the doublet is highest in correspondence to the value of Δ which nullifies the denominator of (4.35), that is

$$\Delta = \frac{1}{2}\left(\frac{1+\omega_{dn}^2}{K\omega_{dn}}\right)^2\left(\sqrt{1+4\left(\frac{K\omega_{dn}}{1+\omega_{dn}^2}\right)^2}-1\right) \tag{4.36}$$

For instance, assuming $K = 2$ and $\omega_{dn} = 2$, Δ is 0.693.

In [PP992] the model was validated for values of Δ in a range from -1 to 0.5, meaning a doublet with its pole and zero spaced by a factor of two.

The time-domain closed-loop step response can be also approximately represented through that of a pure two-pole amplifier. To evaluate the effect of a pole-zero doublet, we calculate the relative deviation of t_p and D in a two-pole amplifier given by (4.23) and (4.24)

$$\frac{\Delta t_p}{t_p} = -\frac{(K-2)(K^2+1)}{K(K-4)}\frac{\omega_{dn}\Delta}{\left(1+\omega_{dn}^2\right)\sqrt{1-\Delta}-K\omega_{dn}\Delta} \tag{4.37}$$

$$\frac{\Delta D}{D} = -2\pi\frac{K(K^2+1)\sqrt{K(4-K)}}{(4-K)^2}\frac{\omega_{dn}\Delta}{\left(1+\omega_{dn}^2\right)\sqrt{1-\Delta}-K\omega_{dn}\Delta} \tag{4.38}$$

For those cases in which Δ is small, such as when a doublet arises from process tolerances in a pole-zero compensation [BAR80], [PP95], relationships (4.37) and (4.38) can be approximated to

$$\frac{\Delta t_p}{t_p} \approx -\frac{(K-2)(K^2+1)}{K(K-4)}\frac{\omega_{dn}}{\left(1+\omega_{dn}^2\right)}\Delta \tag{4.39}$$

$$\frac{\Delta D}{D} \approx -2\pi\frac{K(K^2+1)\sqrt{K(4-K)}}{(4-K)^2}\frac{\omega_{dn}}{\left(1+\omega_{dn}^2\right)}\Delta \tag{4.40}$$

The approximate relationships show that the relative variations of t_p and D are linearly related to the spacing of the doublet. It can be seen that for phase margins greater than 50° (i.e. $K > 1.2$) the variation in D is much greater than that of t_p. Besides, when the zero is lower than the pole, the doublet has the effect of reducing overshoot (both in the frequency and in the time domain).

As discussed in section 4.3, the second pole is often already defined and the compensation task requires setting the dominant pole or, better, the gain-bandwidth product, ω_{GBW}. Relationship (4.29) can be written as

$$\tan(\phi) = \frac{1}{\omega_{GBW}} \frac{\omega_{GBW}^2 (p_2 + p_d \Delta) + p_2 \omega_{dn}^2}{\omega_{GBW}^2 + \omega_d^2 - p_2 p_d \Delta} \tag{4.41}$$

hence, the required ω_{GBW} implies having to solve the following third-order equation

$$\tan(\phi)\omega_{GBW}^3 - (p_2 + p_d \Delta)\omega_{GBW}^2 + \tan(\phi)(\omega_{dn}^2 - p_2 p_d \Delta)\omega_{GBW} - p_2 \omega_{dn}^2 = 0 \tag{4.42}$$

As particular cases, first consider the one where the pole-zero doublet is derived from differential-to-single conversion. In this event doublet spacing, Δ, is exactly equal to -1. By developing (4.42) in Taylor series around the point $\omega_{GBW} = p_2/\tan(\phi)$ truncated to the second term, we get

$$(2p_2 + p_d)\omega_{GBW}^2 + \left[\tan(\phi)(p_2 + p_d)p_d - \frac{3p_2^2}{\tan(\phi)}\right]\omega_{GBW} + \left(\frac{p_2^2}{\tan^2(\phi)} - 2p_d^2\right)p_2 = 0 \tag{4.43}$$

which is sufficiently simple to be solved with pencil-and-paper.

In contrast, when the second pole can be moved to guarantee stability, such as in the design strategy for cascode amplifiers proposed in [MN89], from (4.41) noting that

$$p_d = \frac{\omega_{dn}}{\sqrt{1 - \Delta}} \tag{4.44}$$

we have to set p_2 according to

$$p_2 = \omega_{GBW} \tan(\phi) \left(\frac{1 + \omega_{dn}^2 - \dfrac{\Delta}{\sqrt{1 - \Delta}} \dfrac{\omega_{dn}}{\tan(\phi)}}{1 + \omega_{dn}^2 + \dfrac{\Delta}{\sqrt{1 - \Delta}} \omega_{dn} \tan(\phi)} \right) \tag{4.45}$$

4.4 THREE-POLE FEEDBACK AMPLIFIERS WITH REAL POLES

Some amplifier architectures have three separate poles [P99], one of which must be dominant to allow stability. Consider then the third-order loop-gain transfer function given below

$$T(s) = \frac{T_o}{\left(1 + \dfrac{s}{p_1}\right)\left(1 + \dfrac{s}{p_2}\right)\left(1 + \dfrac{s}{p_3}\right)} \qquad (4.46)$$

The phase margin of the loop gain is equal to (approximating ω_T with ω_{GBW})

$$\phi = \tan^{-1} \frac{p_2}{\omega_{GBW}} + \tan^{-1} \frac{p_3}{\omega_{GBW}} - 90° \qquad (4.47)$$

Since

$$\tan(a + b) = \frac{\tan(a) + \tan(b)}{1 - \tan(a)\tan(b)} \qquad (4.48)$$

and

$$\tan(a + 90°) = -\frac{1}{\tan(a)} \qquad (4.49)$$

relationship (4.47) can be rewritten as

$$\frac{1}{\omega_{GBW}^2} - \tan(\phi)\left(\frac{1}{p_2} + \frac{1}{p_3}\right)\frac{1}{\omega_{GBW}} - \frac{1}{p_2 p_3} = 0 \qquad (4.50)$$

Thus, assuming the non-dominant poles to be definitely set, the required gain-bandwidth for a given phase margin can be achieved from (4.50). In particular, we get

$$\frac{1}{\omega_{GBW}} = \frac{\tan(\phi)}{2} \left(\frac{1}{p_2} + \frac{1}{p_3} \right) \left[1 + \sqrt{1 + \frac{4}{\tan^2(\phi)} \frac{p_2 p_3}{(p_2 + p_3)^2}} \right]$$

$$\text{(4.51)}$$

$$\approx \tan(\phi) \left(\frac{1}{p_2} + \frac{1}{p_3} \right)$$

It is worth noting that the frequency compensation of a three-pole amplifier can be performed following the same procedure as for an equivalent two-pole amplifier with a loop gain given by

$$T(s) \approx T_o \frac{1}{\left(1 + \frac{s}{p_1} \right) \left(1 + \frac{s}{p_{2eq}} \right)} \tag{4.52}$$

The time constant of the equivalent second pole equals the sum of the second and third pole time constants of the three-pole amplifier. In other words, the equivalent pole is

$$p_{2eq} = \frac{p_2 p_3}{p_2 + p_3} \tag{4.53}$$

and the frequency and time-domain behaviour of the closed loop amplifier can be approximated with those developed in section 4.2.

4.5 THREE-POLE FEEDBACK AMPLIFIERS WITH A PAIR OF COMPLEX AND CONJUGATE POLES

Another common situation for a three-pole amplifier is when a dominant pole occurs in conjunction with a pair of complex conjugate poles. We use the symbolism introduced in section 4.3 to express such a loop-gain transfer function

$$T(s) = \frac{T_o}{\left(1 + \frac{s}{p_1} \right) \left(1 + \frac{s}{\omega_{GBWi}} + \frac{s^2}{\omega_{GBWi}^2 K_i} \right)} \tag{4.54}$$

This representation can be particularly useful when the complex poles derive from two indented feedback loops (such as in three-stage amplifiers with nested-Miller[4] compensation). In this case, term ω_{GBWi} is the gain-bandwidth product of the inner loop-gain and K_i is the ratio between the second pole and the gain-bandwidth product in this inner loop. The phase margin of the whole amplifier is given by

$$\phi = 90° - \tan^{-1} \frac{\dfrac{\omega_{GBW}}{\omega_{GBWi}}}{1 - \dfrac{\omega_{GBW}^2}{\omega_{GBWi}^2 K_i}} = \tan^{-1}\left(\frac{1 - \dfrac{\omega_{GBW}^2}{\omega_{GBWi}^2 K_i}}{\dfrac{\omega_{GBW}}{\omega_{GBWi}}} \right) \tag{4.55}$$

hence from (4.55) we get

$$\left(\frac{\omega_{GBWi}}{\omega_{GBW}} \right)^2 - \tan(\phi)\frac{\omega_{GBWi}}{\omega_{GBW}} - \frac{1}{K_i} = 0 \tag{4.56}$$

and we can determine the gain-bandwidth required for a fixed phase margin when the higher poles are fixed

$$\frac{1}{\omega_{GBW}} = \frac{\tan(\phi)}{2}\left[1 + \sqrt{1 + \frac{4}{K_i \tan^2(\phi)}} \right]\frac{1}{\omega_{GBWi}} \tag{4.57}$$

Like the case of three separate poles, now we can define an equivalent second pole and, if the quantity within the square roots is close to one, which means

$$K_i \tan^2(\phi) > 4 \tag{4.58}$$

the equivalent second pole is approximated by ω_{GBWi}. The frequency and time domain behaviour of the closed loop amplifier are hence equal to those of the closed loop amplifier whose open loop transfer function is

$$T(s) \approx T_o \frac{1}{\left(1 + \dfrac{s}{p_1} \right)\left(1 + \dfrac{s}{\omega_{GBWi}} \right)} \tag{4.59}$$

---

[4] See Sec. 5.5 of this book.

4.6 TWO-POLE FEEDBACK AMPLIFIERS WITH A ZERO

Often the loop gain of a feedback amplifier has a zero which can heavily affect the transfer function of the closed-loop amplifier. Indeed, if the loop gain is

$$T(s) = T_o \frac{1 + \dfrac{s}{z}}{\left(1 + \dfrac{s}{p_1}\right)\left(1 + \dfrac{s}{p_2}\right)} \tag{4.60}$$

the closed-loop transfer function exhibits the same zero and is given by

$$G_F(s) = G_{Fo} \frac{1 + \dfrac{s}{z}}{1 + 2\dfrac{\xi}{\omega_o}s + \dfrac{s^2}{\omega_o^2}} \tag{4.61}$$

where in the denominator ω_o is still given by (4.7) and the damping factor is modified with respect to (4.8) according to

$$\xi = \frac{1}{2\sqrt{1 + T_o}}\left(\sqrt{\frac{p_1}{p_2}} + \sqrt{\frac{p_2}{p_1}} + T\frac{\sqrt{p_1 p_2}}{z}\right) \tag{4.62}$$

The phase margin of the feedback amplifier, under the assumption of a dominant-pole behaviour whose pole and zero is higher than the transition frequency, ω_T, is given by

$$\phi = 90° + \arctg\frac{p_2}{\omega_{GBW}} - \arctg\frac{z}{\omega_{GBW}} \tag{4.63}$$

which shows that a negative zero helps stability, but a positive zero can drastically reduce the phase margin. Therefore, during compensation particular care must be taken to avoid or minimise the effect of positive zeros.

Although it is seldom used, the step response of a feedback amplifier with the closed-loop transfer function in (4.61) is

$$x_o(t) = \left\{ 1 - \frac{1}{p_{F1} - p_{F2}} \left[\left(1 - \frac{p_{F1}}{z} \right) p_{F2} e^{-p_{F1}t} - \left(1 - \frac{p_{F2}}{z} \right) p_{F1} e^{-p_{F2}t} \right] \right\} G_{oF} u(t)$$

$$(4.64)$$

where p_{F1} and p_{F2} are the poles given in (4.9). For an underdamped amplifier (4.64) can be expressed more profitably as

$$x_o(t) = \left\{ 1 - \left[H \sin\left(\sqrt{1 - \xi^2}\, \omega_o t \right) + \cos\left(\sqrt{1 - \xi^2}\, \omega_o t \right) \right] e^{-\xi \omega_o t} \right\} G_{oF} u(t) \quad (4.65)$$

where $H = \left[\dfrac{\sqrt{1 - \xi^2}\, \omega_o}{z} + \dfrac{\xi}{\sqrt{1 - \xi^2}} \left(\dfrac{\xi \omega_o}{z} - 1 \right) \right].$

Chapter 5

FREQUENCY COMPENSATION TECHNIQUES

In the previous chapter we demonstrated the necessity, in a feedback network, to achieve an open loop dominant-pole frequency response whose a phase margin is greater than 45° (or $K > 1$). Indeed, this condition not only ensures closed-loop stability but also avoids unacceptably underdamped closed-loop responses. Unfortunately, many amplifiers, and particularly broadbanded amplifiers, earmarked for use as open-loop cells are not characterised by dominant-pole frequency responses. The loop-gain frequency response of these amplifiers must be therefore properly optimised in accordance with standard design practices known as *frequency compensation techniques* [SS91], [GM93], [LS94]. These methods imply the inclusion of compensation *RC* networks in the uncompensated circuit to introduce additional poles or to modify the original loop-gain poles so as to provide a given phase margin.

Referring to Fig. 4.8, it is easily understood that the simplest way to achieve stability is to reduce the loop gain. If the frequency of the poles remain unchanged, the unity-gain frequency is decreased by the same amount as the loop gain reduction and consequently the ratio between the second pole and the gain-bandwidth product is increased. The loop gain can be reduced via the feedback factor f or by decreasing the amplifier open-loop gain. However, neither are practical design choices because changing the loop gain may conflict with closed-loop performance such as gain, accuracy, etc. Moreover, it is worthwhile noting that compensation must be ensured for all the possible feedback configurations. If the feedback factor is not specified, compensation should be performed in the worst-case condition, that corresponds to the unitary feedback (i.e., with the highest loop gain and gain-bandwidth product, $f = 1$ and $T_o = A_o$).

In the following three paragraphs we will study the engineering methods and related tradeoffs underlying the key issue of the frequency compensation for a two-pole open loop transfer function. Of course, the discussion is easily extended to multi-pole functions with two dominant poles. The last two paragraphs deal with the frequency compensation of three-stage amplifiers.

5.1 DOMINANT-POLE COMPENSATION

Let us consider the two-pole amplifier in Fig. 5.1 whose open-loop transfer function is

$$A(s) = \frac{A_o}{\left(1 + \dfrac{s}{p_1}\right)\left(1 + \dfrac{s}{p_2}\right)} \tag{5.1}$$

where

$$A_o = G_{m1} G_{m2} R_1 R_2 \tag{5.2}$$

$$p_1 = \frac{1}{R_1 C_1} \tag{5.3}$$

$$p_2 = \frac{1}{R_2 C_2} \tag{5.4}$$

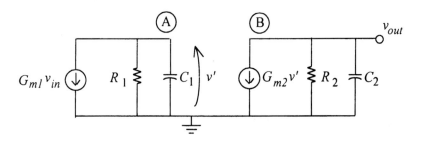

Fig.5.1 Small-signal model of a two-pole amplifier.

The two poles are determined by the parasitic capacitances associated with node A and B. Assuming these poles are widely separated with $p_1 \ll p_2$, the tangent of the phase margin becomes

$$\tan(\phi) = \frac{p_2}{|A_o||p_1} = \frac{1}{|A_o|}\left(\frac{R_2 C_2}{R_1 C_1}\right) \tag{5.5a}$$

Conversely, if $p_2 \ll p_1$ we would have

$$\tan(\phi) = \frac{p_1}{|A_o||p_2} = \frac{1}{|A_o|}\left(\frac{R_1 C_1}{R_2 C_2}\right) \tag{5.5b}$$

To guarantee a phase margin greater than $45°$, $\tan(\phi)$ must be greater than unity. Hence, from (5.5a) and (5.5b), we must ensure that the ratio between the two time constants is in the order of the DC gain. For example, assuming the two equivalent resistances to be equal and a typical gain of 30 one of the capacitances should be more than 30 times the other, to guarantee stability within proper margins.

At this point, the most intuitive way to provide stability is to *add* a capacitance in parallel to C_1 (or C_2), thus setting the dominant pole at the input or the output. If we adopt this strategy, the choice of where to insert the compensation capacitor depends on convenience in terms of lower added capacitance. This simple compensation approach is called dominant-pole compensation, which is rarely used, except in single-stage (cascode) amplifiers, because it requires large compensation capacitors and leads to feedback amplifiers with very low bandwidth. To show the reduction in bandwidth, without loss of generality consider the amplifier as being in unitary feedback and set the dominant pole at the input by adding the compensation capacitor C_C to C_1. Thus (5.5b) turns out to be

$$\tan(\phi) \approx \frac{1}{|A_o|} \frac{R_1(C_1 + C_C)}{R_2 C_2} \tag{5.6}$$

and the dominant pole after compensation, p_{1C}, which defines the open-loop bandwidth

$$p_{1C} \approx \frac{1}{R_1(C_1 + C_C)} \tag{5.7}$$

must be lower than the second pole (which remains unchanged to p_2) reduced by the DC gain times the tangent of the phase margin (always higher than 1)

$$p_{1C} < \frac{p_2}{\tan(\phi)|A_o|} \tag{5.8}$$

To conclude, the bandwidth of the dominant-pole compensated amplifier is defined by the *second pole and the DC open-loop gain*. As we shall see in the next paragraph this condition does not hold for the Miller compensation strategy.

5.2 MILLER (POLE-SPLITTING) COMPENSATION

The well-known *Miller* effect can be efficiently exploited to perform frequency compensation that for this reason is called Miller compensation or pole splitting compensation. To understand its properties and design issues consider the small-signal model in Fig. 5.2, which but for the presence of the interstage capacitance C_r coupling the first and second stage, is equal to the one in Fig. 5.1.

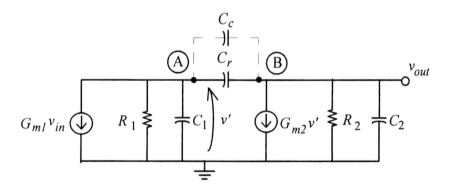

Fig. 5.2. Small-signal model of a two-pole amplifier with interstage capacitance.

Neglecting for the moment capacitor C_C depicted in dashed lines, the subject open-loop transfer function is

$$A(s) = A_o \frac{1 + b_1 s}{1 + a_1 s + a_2 s^2} = A_o \frac{1 + \dfrac{s}{z}}{1 + \left(\dfrac{1}{p_1} + \dfrac{1}{p_2}\right)s + \dfrac{s^2}{p_1 p_2}} \tag{5.9}$$

where the DC gain A_o is still given by (5.2) whose the coefficients are

$$a_1 = R_1C_1 + R_2C_2 + (R_1 + R_2 + R_1G_{m2}R_2)C_r \qquad (5.10)$$

$$a_2 = R_1R_2C_2\left[C_1 + \left(1 + \frac{C_1}{C_2}\right)C_r\right] \qquad (5.11)$$

$$b_1 = -\frac{C_r}{G_{m2}} \qquad (5.12)$$

Thus, assuming the poles are widely separated their approximate expressions become:

$$p_1 \approx \frac{1}{R_2(C_2 + C_r) + R_1[C_1 + (1 + G_{m2}R_2)C_r]} \qquad (5.13)$$

$$p_2 \approx \frac{R_2(C_2 + C_r) + R_1[C_1 + (1 + G_{m2}R_2)C_r]}{R_1R_2C_2\left[C_1 + \left(1 + \frac{C_1}{C_2}\right)C_r\right]} \qquad (5.14)$$

Capacitor C_r provides a path for feedback and for feedforward. The feedforward leakage produces a real zero in the right-half plane (RHP) given by

$$z = -\frac{G_{m2}}{C_r} \qquad (5.15)$$

The effect of this zero is neglected here for simplicity (the zero may be either at a very-high frequency or be compensated with one of the methods described in the next paragraph).

In (5.13) and (5.14), term $(1 + G_{m2}R_2)C_r$ accounts for the Miller effect [MG87]. In practical cases it is the dominant term because capacitance C_r is multiplied by a factor as high as a stage gain, $G_{m2}R_2$. In such cases the expressions of the two poles (5.13) and (5.14) simplify to

$$p_1 \approx \frac{1}{R_1[C_1 + (1 + G_{m2}R_2)C_r]} \qquad (5.16)$$

$$p_2 \approx \frac{C_1 + (1 + G_{m2}R_2)C_r}{R_2 C_2 \left[C_1 + \left(1 + \dfrac{C_1}{C_2}\right)C_r \right]}$$

(5.17)

From equations (5.15) to (5.17) the pole splitting due to Miller effect becomes apparent. In fact, an increase in the internal feedback capacitance, C_r, shifts the dominant pole and the second pole to a lower and higher frequency, respectively (and also decreases the RHP zero). For this purpose, to improve the separation of the two poles it is very efficient to multiply C_r.

Thus, pole-splitting compensation entails connecting a capacitor between two phase inverting nodes of the open-loop amplifier. With reference to the equivalent circuit in Fig. 5.2, the electrical impact of this additional element is the replacement of the internal interstage capacitance, C_r, by the capacitance sum, $C_C + C_r$.

Letting $C_P = C_C + C_r$, (5.16) and (5.17) can be further simplified to

$$p_{1C} \approx \frac{1}{R_1 G_{m2} R_2 C_p}$$

(5.18)

$$p_{2C} \approx \frac{G_{m2}}{(C_2 + C_1)}$$

(5.19)

where capacitance C_p is usually significantly larger than C_r and has also been assumed to be greater than either C_1 or C_2. Note that the value of the compensated second pole given by (5.19) encounters an intuitive justification. In fact, at the frequency at which it occurs (i.e. after the transition frequency or equivalently the gain-bandwidth product), capacitance C_p can be considered as short-circuited. Hence, the input and the output of the voltage-controlled current-source are shorted, C_2 and C_1 are in parallel and the equivalent resistance seen at their terminals is approximately $1/G_{m2}$.

In contrast, the expression of the zero (5.15) becomes

$$z_C = -\frac{G_{m2}}{C_r + C_C}$$

(5.20)

Although z_C can exert a significant influence on the high-frequency response of the compensated amplifier, the following discussion presumes tacitly that $z_C > p_{2C}$. Hence, the gain-bandwidth product and the phase margin are

$$\omega_{GBW} = \frac{G_{m1}}{C_r + C_C} \tag{5.21}$$

$$\tan(\phi) = \frac{G_{m2}}{G_{m1}} \frac{C_r + C_C}{C_2 + C_1} \tag{5.22}$$

and the required compensation capacitance must be set according to

$$C_C = \frac{G_{m1}}{G_{m2}} \tan(\phi)(C_2 + C_1) - C_r \tag{5.23}$$

For a fixed phase margin C_C is proportional to the ratio between the transconductance of the first and second stage. Moreover, it is proportional to the total (input-output) capacitance. Note also that for a given DC gain and phase margin, the gain-bandwidth product is set by the frequency of second pole. Consequently, to compare the Miller and dominant-pole compensations we can compare only the second poles, and it is apparent that the second pole resulting from the Miller compensation is much higher (due to pole-splitting) than that of a dominant-pole compensated amplifier. In addition, Miller magnification allows us to use lower capacitance values.

For these reasons the Miller compensation technique is extensively used to design IC amplifiers. Compensation capacitor C_C can be fabricated as a part of the amplifier (in this case the amplifier is said to be *internally compensated*) or can be externally applied to pins reserved for this purpose on the (*uncompensated*) opamp package.

By comparing (5.20) and (5.21), we find that we can neglect the right-half plane zero when the transconductance gain of the second stage is much higher than that of the first stage. This condition is seldom satisfied in CMOS transconductance amplifiers and especially when low-power dissipation is required, so that a specific strategy to compensate the zero must be applied.

5.3 COMPENSATION OF THE MILLER RHP ZERO

In the previous paragraph we showed that the pole-splitting technique is a convenient vehicle for achieving the desired pole separation in an open-loop phase-inverting amplifier. Unfortunately, (5.20) indicates that the larger the C_C, the lower the RHP zero. In bipolar technologies the transconductance G_{m2} is invariably large enough to ensure that the frequency of the zero is greater than the compensated unity-gain frequency, thereby rendering the impact of z_c on the compensated frequency response inconsequential. But for MOS and CMOS technologies, the transconductance is small, and as a result, the effects of the RHP zero evidenced in the forward transfer function of a phase inverting amplifier may not be negligible. When the transmission zero is significant, its primary effect is to incur excess phase lag (phase lag in addition to that produced by the two open-loop poles), while prohibiting a uniform 20 dB-per-decade frequency response roll-off rate at high frequencies. The stability problems caused by the resultant deterioration in phase margin justifies the implementation of compensation techniques that neutralise the effects of the RHP zero.

Various compensation schemes have been proposed for two-stage MOS opamps. They are based on the concept of breaking the forward path through the compensation capacitor by using active or passive components. The first of these was applied in a NMOS opamp [TG76] and then in a CMOS opamp [SHG78]. It breaks the forward path by introducing a voltage buffer in the compensation branch. Next, a compensation technique was proposed which uses a nulling resistor in series with the compensation capacitor [A83]. Another solution works like the former but uses a current buffer to break the forward path [A83]. Finally, both current and voltage buffers can be adopted to compensate the right half-plane zero [MT90].

5.3.1 Nulling Resistor

The most widely used compensation technique is the one based on the nulling resistor. It entails the incorporation of a resistor, R_C, in series with the Miller compensation capacitor as shown in Fig. 5.3.

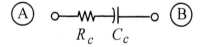

Fig. 5.3. Compensation network using a nulling resistor.

The popularity of this scheme stems from the fact that it can be implemented monolithically with a MOS transistor biased in its triode regime (which approximates a linear resistor). Moreover, its highpass nature does not reduce the low-frequency dynamic range of the uncompensated configuration. By using this compensation branch in the equivalent circuit in Fig. 5.2, and neglecting capacitance C_r (usually much lower than C_c), the zero is now at frequency

$$z_r = -\frac{1}{\left(\dfrac{1}{G_{m2}} - R_C\right)C_C}$$ (5.24)

and is moved to infinite frequency by setting R_C equal to $1/G_{m2}$. Thus, the RHP zero originally ignored in the process of arriving at the pole-splitting results has effectively been eliminated.

If R_C is greater than $1/G_{m2}$, a left-hand zero is created because z_r becomes positive. Ideally, this zero can be exploited to offset or even cancel the effects of the second compensated pole, thereby leading to an open-loop amplifier with an increased gain-bandwidth product as first proposed in [BAR80].

By imposing the condition

$$\frac{G_{m2}}{C_2} = \frac{G_{m2}}{(G_{m2}R_C - 1)C_C}$$ (5.25)

a new second pole arises which is given by $1/(R_C C_C)$, as can be found by directly analysing the equivalent circuit. This pole does not depend on the load capacitance. However, this optimised approach has a quite worse ω_{GBW} than the other optimised compensation strategies for equal power consumption and area of the amplifier including the compensation network (i.e., global transconductance in the amplifier) as demonstrated in [PP95] and [PP97].

5.3.2 Voltage Buffer

Figure 5.4 shows the compensation branch with a voltage buffer. It eliminates feedforward through the compensation capacitance C_C.

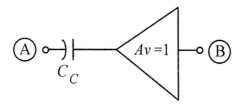

Fig. 5.4. Compensation network with an ideal voltage buffer.

Unfortunately, unlike the passive compensation strategy discussed above, the buffer utilised attenuates the achievable output swing of the amplifier. The adoption of an ideal voltage buffer (i.e., with infinitely large input impedance, zero output impedance, and unitary gain) gives the same dominant pole as in (5.18) and the same second pole as in (5.19) without depending on capacitance C_1. But by eliminating capacitive feedforward, the troublesome RHP zero incurred by the internal interstage capacitance, C_r, is not decreased by the compensation element C_C. In other words, the effective feedback capacitance is C_p, while the feedforward capacitance is C_r.

The foregoing discussion presumes an ideal voltage buffer. Practical buffers have small, but not zero, output impedance and large, but not infinite, input impedance (see Fig. 5.5).

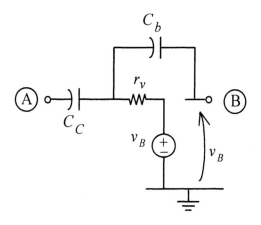

Fig. 5.5. Compensation network with a real voltage buffer.

The resistive component of the buffer output impedance, r_v, establishes a left-half plane zero with capacitance C_C. As for the case with the nulling resistor, this zero can be exploited to increase the amplifier gain-bandwidth [PP95]. Following this last compensation strategy and with some approximations, the poles and zeros become [PP95]

$$p_{1C} \approx \frac{1}{R_1 G_{m2} R_2 C_C} \tag{5.26}$$

$$p_{2C} \approx \frac{G_{m2}}{C_2 + C_b G_{m2} r_v} \tag{5.27}$$

$$p_{3C} \approx \frac{1}{(C_1 + C_b) r_v} \tag{5.28}$$

$$z_{1C} = \frac{1}{C_C r_v} \qquad (5.29)$$

$$z_{2C} = -\frac{G_{m2}}{C_b} \qquad (5.30)$$

Now the right-half plane zero, z_{2C}, is placed at a very high frequency and can be neglected. Moreover, as proposed in [AH87] and developed in [PP95], a pole-zero compensation can be performed to increase the gain bandwidth product. In particular, we can properly design the voltage buffer to ensure the output resistance is equal to

$$r_v = \frac{C_2}{G_{m2}(C_C - C_b)} \approx \frac{C_2}{G_{m2}C_C} \qquad (5.31)$$

which sets $z_{1C} = p_{2C}$. The new second pole is now the old third pole in (5.28) which by using (5.31) becomes

$$p_{2CN} = p_{3C} = \frac{G_{m2}}{C_2} \frac{C_C}{C_1 + C_b} \qquad (5.32)$$

The phase margin is given by

$$\tan(\phi) = \frac{G_{m2}}{G_{m1}} \frac{C_C^2}{C_2(C_1 + C_b)} \qquad (5.33)$$

which yields the required compensation capacitance

$$C_C = \sqrt{\tan(\phi) \frac{G_{m1}}{G_{m2}} (C_1 + C_b) C_2} \qquad (5.34)$$

After substituting the value found in the gain bandwidth product we get

$$\omega_{GBW} = \frac{G_{m1}}{\sqrt{\tan(\phi) \frac{G_{m1}}{G_{m2}} (C_1 + C_b) C_2}} \qquad (5.35)$$

The resulting ω_{GBW} has a higher value than that given by (5.21), and is inversely dependent on the geometric media of $C_1 + C_b$ and C_2.

5.3.3 Current Buffer

Consider now the ideal compensation branch using the current buffer depicted in Fig. 5.6. This solution is very efficient both for the gain-bandwidth [C93], [RK95] and PSRR performance [A83], [RC84], [SS90], [SGG91]. Moreover, it does not have the drawback exhibited by the voltage buffer of reducing the amplifier output swing.

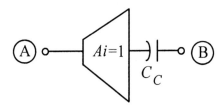

Fig. 5.6. Compensation network with an ideal current buffer.

With an ideal current buffer, the second pole is given by

$$p_{2c} = \frac{G_{m2}}{C_1\left(1 + \dfrac{C_2}{C_C}\right)} \tag{5.36}$$

and the phase margin is by

$$\tan(\phi) = \frac{G_{m2}}{G_{m1}} \frac{C_C^2}{C_1(C_C + C_2)} \tag{5.37}$$

By solving for the compensation capacitance we found

$$C_C = \tan(\phi)\frac{G_{m1}}{2G_{m2}}\left(1 + \sqrt{1 + \frac{4G_{m2}}{\tan(\phi)G_{m1}}\frac{C_2}{C_1}}\right)C_1 \approx$$

$$\approx \tan(\phi)\frac{G_{m1}}{2G_{m2}}C_1 + \sqrt{\tan(\phi)\frac{G_{m1}}{G_{m2}}C_1C_2} \tag{5.38}$$

Generally the output capacitance, C_2, is much higher than the inner capacitance, C_1, and relationship (5.38) can be further simplified to

$$C_C \approx \sqrt{\tan(\phi)\frac{G_{m1}}{G_{m2}}C_1C_2} \tag{5.39}$$

Hence, for a given phase margin, the required compensation capacitance is slightly lower than the value required by the optimised compensation with voltage buffer in (5.34), while the resulting gain bandwidth product is slightly higher.

However, compensation with a real current buffer (and specifically, with finite input resistance) is not as straightforward as the other compensation approaches. As shown in reference [PP97], to achieve compensation we have to guarantee that the input resistance of the current buffer, r_c, is equal to or lower than half $1/G_{m1}$. Moreover, the condition

$$r_c = \frac{1}{2G_{m1}} \tag{5.40}$$

represents an optimum to maximise the gain-bandwidth product. Under condition (5.40) the required compensation capacitor is

$$C_C \approx h\frac{G_{m1}}{2G_{m2}}C_1 + \sqrt{\left(h + \frac{1}{2}\right)\frac{G_{m1}}{G_{m2}}C_1C_2} \tag{5.41}$$

where

$$h = \frac{2\tan(\phi)-1}{2+\tan(\phi)} \tag{5.42}$$

Usually, relationship (5.41) can be further approximated

$$C_C \approx \sqrt{\left(h+\frac{1}{2}\right)\frac{G_{m1}}{G_{m2}}C_1C_2} \qquad (5.43)$$

and the gain bandwidth product results as

$$\omega_{GBW} \approx \frac{G_{m1}}{\sqrt{\left(h+\frac{1}{2}\right)\frac{G_{m1}}{G_{m2}}C_1C_2}} \qquad (5.44)$$

For practical phase margin values, the gain bandwidth product in (5.44) is even higher than that obtained with a ideal current buffer. It is also higher than the one obtained using a real voltage buffer. However, compensation with a real current buffer is a less efficient strategy because, as demonstrated in [PP97], it needs more area and/or power for equal gain bandwidth product than compensation based on a real voltage buffer.

5.4 NESTED MILLER COMPENSATION

The compensation of multistage amplifiers (i.e., with a number of gain stages higher than two) requires iteration of the simple Miller compensation described previously [C78], [C821], [C96], [HL85], [EH95]. Typically, three- and even four-stage amplifiers are found in CMOS implementations including an output power stage for driving heavy off-chip loads [C822], [OA90], [PD90], [TGC90], [CN91], [PNC93]. Moreover, given the decrease in supply voltages, *cascoding* is not a suitable technique for IC applications demanding both high gain and swing. Hence, *cascading* three or more simple stages is the only viable option. Consequently, multistage amplifiers and their frequency compensation issues have become increasingly important in modern microelectronics [FH91], [EH92], [NG93], [PPS99], [GPP00]. In the following we will discuss in detail compensation of three-stage amplifiers, which are the most common architectures, but the results obtained can also be adapted (although often not very directly) to architectures with a higher number of stages.

5.4.1 General Features

Among the possible ways of exploiting Miller compensation in multistage amplifiers, the so-called *nested Miller* (NM) compensation is one of the most widely used. It can be utilised when only the final gain stage is voltage-inverting.

The small-signal equivalent circuit of a three-stage amplifier including nested Miller compensation is depicted in Fig. 5.7. Parameters G_{mi} and R_i are the i-th stage transconductance and output resistance, respectively. Capacitors C_i represent the equivalent capacitance at the output of each stage, C_{Ci} are the compensation capacitors, and C_L is the equivalent load capacitor.

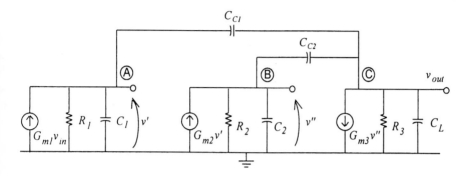

Fig. 5.7. Small-signal model of a three-stage NM-compensated amplifier.

In the following we neglect the effects of the parasitic capacitances since they are generally one order of magnitude lower than the compensation capacitances. Neglecting second-order terms, the open-loop transfer function of the circuit in Fig. 5.7 is expressed by

$$A(s) = A_o \frac{1 - \dfrac{C_{C2}}{G_{m3}}s - \dfrac{C_{C1}C_{C2}}{G_{m2}G_{m3}}s^2}{\left(1 + \dfrac{s}{p_1}\right)\left[1 + \left(\dfrac{1}{G_{m2}} - \dfrac{1}{G_{m3}}\right)C_{C2}s + \dfrac{C_{C2}C_L}{G_{m2}G_{m3}}s^2\right]} \qquad (5.45)$$

where A_o is the DC open-loop gain equal to

$$A_o = G_{m1}G_{m2}G_{m3}R_1R_2R_3 \qquad (5.46)$$

and p_1 is the frequency of the dominant pole

$$p_1 \approx \frac{1}{R_1 G_{m2} R_2 G_{m3} R_3 C_{C1}} \qquad (5.47)$$

Hence, the gain-bandwidth product, ω_{GBW}, of the amplifier is equal to

$$\omega_{GBW} = \frac{G_{m1}}{C_{C1}} \tag{5.48}$$

Equation (5.45), in addition to a dominant pole, also includes two other higher poles and two zeros. Moreover, since the coefficients of the s and s^2 terms in the numerator are both negative, a RHP zero is created, which is located at a lower frequency than the other LHP zero. In analogy to the discussion of the previous paragraph, using voltage followers or current followers can nominally eliminate both zeros. Another solution is the *multipath* Miller approach proposed in [YES97] that, according to Fig. 5.8, provides a zero cancellation due to the effect described in [EH95]. In brief, the forward path contribution is ideally nullified by setting G_{mc} equal to G_{m1}

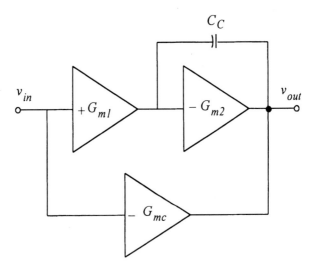

Fig. 5.8. Basic module for *multipath* nested-Miller zero cancellation.

When using any of these techniques, or in the case of a very large G_{m3}, such as in power amplifiers (whose output stage is biased with large quiescent currents and is realised with large devices), relationship (5.45) simplifies to

$$A(s) = A_o \frac{1}{\left(1 + \dfrac{s}{p_1}\right)\left(1 + \dfrac{C_{C2}}{G_{m2}} s + \dfrac{C_{C2} C_L}{G_{m2} G_{m3}} s^2\right)} \tag{5.49}$$

Equation (5.49) allows an interesting interpretation of the compensation process. We will show that assuming a dominant-pole frequency response, the second and third stage can be considered as closed in a unity-gain feedback configuration by capacitor C_{C1}, acting as a short circuit for frequencies above ω_{GBW}.

Consider now the open-loop gain of the second and third stage alone (which we also refer to as the *inner* amplifier), $A_i(s)$, its DC gain, the dominant pole p_{1i} due to the Miller effect on C_{C2}, and the second pole p_{2i} at the output terminal. They are given by

$$A_{oi} = G_{m2} G_{m3} R_2 R_3 \tag{5.50}$$

$$p_{1i} \approx \frac{1}{R_2 G_{m3} R_3 C_{C1}} \tag{5.51}$$

$$p_{2i} \approx \frac{G_{m3}}{C_L} \tag{5.52}$$

If now we assume $A_i(s)$ in unity-gain feedback connection, the resulting closed-loop transfer function is characterised by exactly the same second-order polynomial as in the denominator of (5.49). This consideration justifies the representation utilised in equation (4.54) and allows, in turn, the straightforward compensation technique discussed below.

For a well designed (i.e., with appropriate stability margins) inner amplifier, the second pole p_{2i} must be located well beyond the unity-gain frequency ω_{Ti}, which, under the dominant-pole behaviour assumption, is approximately equal to ω_{GBWi} and given by

$$\omega_{GBWi} \approx \frac{G_{m2}}{C_{C2}} \tag{5.53}$$

In order to avoid overshoot in the module of the inner amplifier frequency response, a proper ratio, K_i, between p_{2i} and ω_{GBWi} has to be set as described in paragraph 4.5. A fairly optimum value of K_i is 2 (i.e. an inner phase margin of about 64°) which is the minimum value guaranteeing monotonic behaviour in the frequency response module. This leads to the expression of capacitor C_{C2},

$$C_{C2} = 2 \frac{G_{m2}}{G_{m3}} C_L \tag{5.54}$$

In other words, we have an external feedback loop through C_{C1} and an inner one through C_{C2}. The stability of the inner loop must first be established so that we can proceed to the external one. Any design attempt not providing a proper phase margin for the inner loop would inevitably require an extremely high value of C_{C1} or even not achieve stability at all.

Now we return to the frequency response of the whole open-loop amplifier, which can be rewritten as in (4.54) and is here reported for clarity

$$A_o(s) = A_o \frac{1}{1 + \dfrac{s}{p_1}} \frac{1}{1 + \dfrac{s}{\omega_{GBWi}} + \dfrac{s^2}{2\omega_{GBWi}^2}} \tag{5.55}$$

Evaluation of the phase margin yields (see (4.55))

$$\phi = 90° - \tan^{-1} \frac{\dfrac{\omega_{GBW}}{\omega_{GBWi}}}{1 - \dfrac{\omega_{GBW}^2}{2\omega_{GBWi}^2}} = \tan^{-1}\left(\frac{1 - \dfrac{\omega_{GBW}^2}{2\omega_{GBWi}^2}}{\dfrac{\omega_{GBW}}{\omega_{GBWi}}} \right) \tag{5.56}$$

Solving (5.56) for ω_T/ω_{Ti} and combining with (5.48) and (5.53) gives the expression of capacitance C_{C1} as a function of the required phase margin

$$C_{C1} = \left[\tan(\phi) + \sqrt{\tan^2(\phi) + 2} \right] \frac{G_{m1}}{G_{m3}} C_L \tag{5.57}$$

Equations (5.54) and (5.57), are very similar to those in [EH95], where a third-order Butterworth frequency response in unity-gain configuration is assumed. However, (5.57) is more general because allows to set compensation capacitor C_{C1} for the desired phase margin.

5.4.2 RHP Cancellation with Nulling Resistors

Now we extend the considerations on the nulling resistor network reported in 5.3.1, to the three-stage nested-Miller compensated amplifier. Figure 5.9 illustrates the RC compensation network which includes two nulling resistors R_{C1} and R_{C2}, to be used in the amplifier of Fig. 5.2.

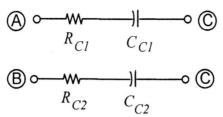

Fig. 5.9. Conventional compensation network using two nulling resistors.

With the introduction of these two resistors the open-loop gain given in (5.45) changes to

$$A(s) = A_o \frac{1 + \left[R_{C1}C_{C1} + \left(R_{C2} - \frac{1}{G_{m3}} \right)C_{C2} \right]s - \frac{1 + (1 - G_{m3}R_{C2})G_{m2}R_{C1}}{G_{m2}G_{m3}}C_{C1}C_{C2}s^2}{\left(1 + \frac{s}{p_1} \right)\left[1 + \left(R_{C2} + \frac{1}{G_{m2}} - \frac{1}{G_{m3}} \right)C_{C2}s + \frac{C_{C2}C_L}{G_{m2}G_{m3}}s^2 \right]} \tag{5.58}$$

Observe that only R_{C2} modifies the denominator because R_{C2} changes the zero of the inner amplifier. It is also clear that the numerator of (5.58) is greatly different from that of (5.45) and now depends on R_{C1} and R_{C2}. By inspection of (5.58), it is possible to nullify the s^2 term and to make the s term positive by choosing

$$R_{C2} = \frac{1}{G_{m2}G_{m3}R_{C1}} + \frac{1}{G_{m3}} \tag{5.59}$$

In this manner, the residual LHP zero can be exploited to increase the phase margin. However, we shall not further develop this approach because better ones have been elaborated.

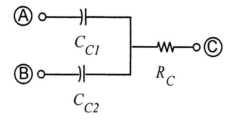

Fig. 5.10. Compensation network using a single nulling resistor.

A simpler technique based on a single nulling resistor and illustrated in Fig. 5.10 was proposed in [LM99]. When applied to the amplifier in Fig. 5.7, it gives the following loop-gain expression

$$A(s) = A_o \frac{1 + \left[R_C C_{C1} + \left(R_C - \dfrac{1}{G_{m3}} \right) C_{C2} \right] s + \dfrac{G_{m3} R_C - 1}{G_{m2} G_{m3}} C_{C1} C_{C2} s^2}{\left(1 + \dfrac{s}{p_1} \right) \left[1 + \left(\dfrac{1}{G_{m2}} - \dfrac{1}{G_{m3}} \right) C_{C2} s + \dfrac{1 - G_{m2} R_C}{G_{m2} G_{m3}} C_{C2} C_L s^2 \right]} \qquad (5.60)$$

In this case the s^2 term in the numerator can be simply set equal to zero by choosing

$$R_C = \frac{1}{G_{m3}} \qquad (5.61)$$

and the loop-gain only has a negative zero which can be used to increase the phase margin.

Now equation (5.54) cannot be used, but the same procedure can still be adopted to achieve simple new equations for C_{C1} and C_{C2}, for a given value of phase margin. After having substituted (5.61) in (5.60) and assuming $K_i = 2$, we can set C_{C2} and evaluate the phase margin

$$C_{C2} = 2 \frac{G_{m2}}{G_{m3} - G_{m2}} C_L \qquad (5.62)$$

$$\phi = 90° - \tan^{-1} \frac{\dfrac{\omega_{GBW}}{\omega_{GBWi}}}{1 - \dfrac{\omega_{GBW}^2}{2\omega_{GBWi}^2}} + \tan^{-1} \frac{\omega_{GBW}}{z} =$$

$$(5.63)$$

$$= \tan^{-1} \left\{ \frac{2\omega_{GBWi} \left(\dfrac{\omega_{GBW}}{\omega_{GBWi}} \right)^2 + z \left[2 - \left(\dfrac{\omega_{GBW}}{\omega_{GBWi}} \right)^2 \right]}{\dfrac{\omega_{GBW}}{\omega_{GBWi}} \left\{ \omega_{GBWi} \left[\left(\dfrac{\omega_{GBW}}{\omega_{GBWi}} \right)^2 - 2 \right] + 2z \right\}} \right\}$$

where z is the zero $1/R_C C_{C1}$ and ω_{GBWi}, comparing (5.55) with (5.60) and using (5.62), is $\frac{1}{2}G_{m3}/C_L$. Solving (5.63) for $\omega_{GBWi}/\omega_{GBW}$ and combining with $\omega_{GBW} = G_{m1}/C_{C1}$ and (5.61) gives the value of capacitance C_{C1}

$$C_{C1} = \frac{\left[\sqrt{\chi^2\left(2\tan^2(\phi)+1\right)+2\chi\tan(\phi)+2+\tan^2(\phi)}+\tan(\phi)-\chi\right]}{1+\chi\tan(\phi)}\chi C_L$$

(5.64)

where

$$\chi = \frac{G_{m1}}{G_{m3}}$$

(5.65)

By considering that χ is lower than $\tan(\phi)$ for the phase margin of interest (i.e., for $\phi \geq 60°$), the above equation can be approximated as

$$C_{C1} \approx \frac{\tan(\phi)+\sqrt{2+\tan^2(\phi)+2\chi\tan(\phi)}}{1+\chi\tan(\phi)}\chi C_L$$

(5.66)

which is more suitable for pencil and paper design, and provides the same results as (5.57) for $\chi \ll 1$. Compared to (5.57), relationship (5.66) gives lower values of C_{C1} for the same phase margin.

It is interesting to note that we assumed no constraint for transconductances except $G_{m3} > G_{m2}$, otherwise C_{C2} in (5.62) would be negative. This allows the power consumption to be optimised since low quiescent currents can be used and, perhaps more importantly, we are free to choose the input and output transconductances G_{m1} and G_{m3}. Unfortunately, like for classic NM compensation, this method still requires large compensation capacitors for heavy capacitive loads. For instance, if $g_{m3} = 4g_{m1}$ and for a target phase margin of 70°, the required compensation capacitor C_{C1} equals $0.9C_L$.

An alternative and efficient compensation technique is based on the compensation network shown in Fig. 5.11 [PP02]. The previous single-resistor compensation network is here modified by adding another resistor, R_{C2}, in series with capacitor C_{C2}. Although this change may appear of marginal significance, it turns out to be very attractive because it allows pole-zero compensation to be achieved by using reduced compensation capacitor values. This in turn leads to an improvement in terms of gain-bandwidth product, slew-rate and settling time.

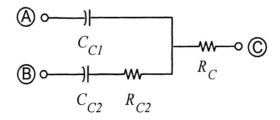

Fig. 5.11. Alternative two-resistor compensation network.

The transfer function of the amplifier in Fig. 5.7, using the compensation network in Fig. 5.11 becomes

$$A(s) = A_o \frac{1 + \left[R_C C_{C1} + \left(R_{C2} + R_C - \dfrac{1}{G_{m3}} \right)C_{C2} \right]s + \dfrac{(1 + G_{m2}R_{C2})G_{m3}R_C - 1}{G_{m2}G_{m3}}C_{C1}C_{C2}s^2}{\left(1 + \dfrac{s}{p_1}\right)\left[1 + \left(R_{C2} + \dfrac{1}{G_{m2}} - \dfrac{1}{G_{m3}} \right)C_{C2}s + \dfrac{1 - G_{m2}R_C}{G_{m2}G_{m3}}C_{C2}C_L s^2 \right]} \tag{5.67}$$

The above shows that the zeros can both be made negative and their values adjusted to exactly cancel the two higher poles. Hence, by setting

$$R_C = \frac{1}{G_{m3}} \tag{5.68}$$

and equating the coefficients of the second-order polynomials we get

$$C_{C1} = \left(\frac{G_{m3}}{G_{m2}} - 1 \right)C_{C2} \tag{5.69}$$

$$R_{C2} = \frac{1}{G_{m3}} \frac{C_L}{C_{C2}} \tag{5.70}$$

The transfer function of the amplifier in (5.67) now has a single pole. This means that a suitable value of C_{C1} can be chosen to maximise the gain-bandwidth product, allowing it to reach the same order of magnitude as an optimised two-stage Miller-compensated amplifier. Again G_{m3} must be higher than G_{m2}, so that the compensation elements will be positive. Moreover, it is worth noting that relations (5.68)-(5.70) are independent of G_{m1} and, ideally, the compensation capacitors are also independent of the load capacitor.

Of all the possible solutions that reduce (5.67) to a single-pole function, the one chosen also has the property of providing an inherent pole-zero cancellation for the (open-loop) transfer function of the amplifier containing only the second and third stage. Indeed, by denoting their second pole and (negative) zero as $\widetilde{p}_2$ and $\widetilde{z}$, respectively, these are given by

$$\widetilde{p}_2 = \frac{G_{m3}}{C_L} \tag{5.71}$$

$$\widetilde{z} = \frac{1}{\left(R_{C2} + R_C - \frac{1}{G_{m3}}\right)C_{C2}} \tag{5.72}$$

whose expressions perfectly match if equations (5.68)-(5.70) are used. However, note that the *inner* amplifier, which is closed in the feedback loop by capacitor C_{C1}, is now comprised between the input of the second stage and the common node of R_{C2} and R_C. Therefore, according to our design methodology, we firstly have to check the stability of this feedback loop. The open-loop transfer function of the inner amplifier is

$$A_i(s) = A_{io} \frac{1 + \left(R_{C2} - \frac{1}{G_{m3}}\right)C_{C2}s - \frac{R_C C_{C2} C_L}{G_{m3}}s^2}{1 + G_{m3} R_2 R_3 C_{C2}s + R_2 R_3 C_{C2} C_L s^2} \tag{5.73}$$

where A_{io} is given by (5.50). From (5.73) and using (5.68) and (5.70) we derive the expressions of the unity-gain frequency and those of the second pole and zeros of the inner amplifier

$$\omega_{GBWi} = \frac{G_{m2}}{C_{C2}} \tag{5.74}$$

$$p_{2i} \approx \frac{G_{m3}}{C_L} \tag{5.75}$$

$$z_{1i} \approx \frac{G_{m3}}{C_L - C_{C2}} \tag{5.76}$$

$$z_{2i} \approx -\left(\frac{C_L}{C_{C2}} - 1\right)\frac{G_{m3}}{C_L} \tag{5.77}$$

indicating that the second pole and the first zero remain very close provided that $C_{C2} \ll C_L$. In this case, the second (RHP) zero tends to G_{m3}/C_{C2} and must be higher than the unity-gain frequency given in (5.74). to ensure stability. Setting the inner phase margin greater than 64° yields $G_{m3} > 2G_{m2}$. The above relation establishes a lower limit for the ratio between G_{m3} and G_{m2}. Under this condition, any value of $C_{C2} \ll C_L$ ideally ensures the stability of the inner amplifier. A minimum usable value for C_{C2} exists in reality. Compensation capacitors must be greater than the parasitic capacitances at the high-impedance nodes to be valid for development. Besides, and usually more importantly, slew-rate considerations posit the fundamental limit for the minimum value of C_{C2} [PP02], [PPP01].

5.5 REVERSED NESTED MILLER COMPENSATION

When the amplifier is made up of three gain stages and the inner stage is the only inverting one, reversed nested Miller compensation (RNMC) becomes the most suitable technique [EH95].

5.5.1 General Features

Figure 5.12 shows a three-stage amplifier small-signal circuit including reversed nested Miller compensation performed by capacitors C_{C1} and C_{C2}. As usual, parameters G_{mi} and R_i are the i-th stage transconductance and output resistance, respectively. Capacitors C_i represent the equivalent capacitance at the output of each stage, while C_L is the equivalent load capacitor. Since capacitor C_{C2} has no connection with the load capacitor (but only with parasitic capacitor C_2), inner loop stability is virtually achieved for all practical C_{C2} values, and will not be examined. For the same reason, this technique has an inherent bandwidth advantage over other multistage compensation approaches based on the Miller effect.

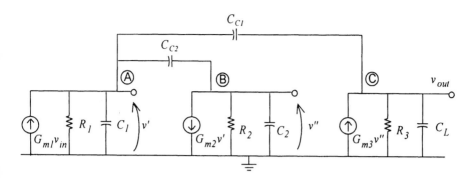

Fig. 5.12. Small-signal model of a three-stage amplifier with reversed nested Miller compensation.

Neglecting second-order terms, the open-loop gain of the circuit in Fig. 5.12 is given by

$$A(s) = A_o \frac{1 - \left(\dfrac{C_{C2}}{G_{m2}} + \dfrac{C_{C1}}{G_{m2}G_{m3}R_2} \right)s - \dfrac{C_{C1}C_{C2}}{G_{m2}G_{m3}}s^2}{\left(1 + \dfrac{s}{p_1}\right)\left[1 + \left(\dfrac{C_{C2}C_L}{G_{m3}C_{C1}} - \dfrac{C_{C2}}{G_{m2}} + \dfrac{C_{C2}}{G_{m3}} \right)s + \dfrac{C_{C2}C_L}{G_{m2}G_{m3}}s^2\right]} \qquad (5.78)$$

where A_o is the DC open-loop gain equal to $G_{m1}R_1G_{m2}R_2G_{m3}R_3$ and p_1 is the dominant pole due to compensation capacitor C_{C1}. Therefore, the dominant pole and the gain-bandwidth product are equal to those of the nested Miller compensation in (5.47) and (5.48), respectively. Moreover, again as for the NMC, we have two other higher poles (usually complex and conjugates) and two zeros, the lower one on the right-half plane and the other on the left-half plane.

Unlike in the NMC, large values of G_{m3} do not facilitate the task of compensation. If G_{m3} is much higher than G_{m2}, except when considering parasitic capacitances, a pole-zero cancellation occurs which modifies (5.78) into a single pole transfer function. But the pole and the zero involved in this compensation are positive, a condition which is critical for stability. Therefore, we must provide viable compensation procedures also for large values of G_{m3}.

First observe that both zeros can be eliminated by using two voltage or current buffers in series with compensation capacitors to break the forward paths, as illustrated in Fig. 5.13 and 5.14, respectively.

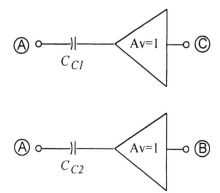

Fig. 5.13. Compensation network with voltage buffers.

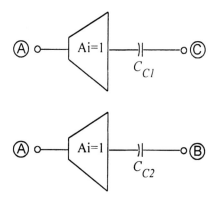

Fig. 5.14. Compensation network with current buffers.

In these two cases (5.78) respectively becomes

$$A(s) = A_o \frac{1}{\left(1 + \dfrac{s}{p_1}\right)\left(1 + \dfrac{C_{C2}C_L}{G_{m3}C_{C1}}s\right)} \tag{5.79}$$

$$A(s) = A_o \frac{1}{\left(1 + \dfrac{s}{p_1}\right)\left[1 + \dfrac{C_{C2}(C_{C1}+C_L)}{G_{m3}C_{C1}}s\right]} \tag{5.80}$$

It is worth noting that the above expressions have exactly two poles thanks to the action of the ideal buffers. Both second poles are also negative. The second pole in (5.79) can be simply interpreted by analysing the circuit in Fig. 5.15, where the inner amplifier, A_2, through the capacitive network, acts as a G_{m3} multiplier by a factor equal in module to C_{C1}/C_{C2}. The same considerations hold for the second pole of (5.80). The only difference is that C_{C1} and C_L are now in parallel.

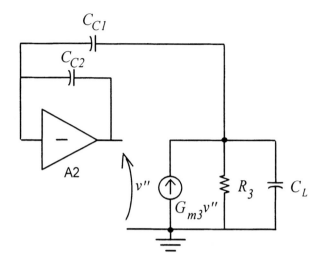

Fig. 5.15. Circuit model for the evaluation of the second pole.

The specified phase margin for the two cases respectively is given by

$$\tan(\phi) = \frac{G_{m3}}{G_{m1}} \frac{C_{C1}^2}{C_{C2} C_L} \tag{5.81}$$

$$\tan(\phi) = \frac{G_{m3}}{G_{m1}} \frac{C_{C1}^2}{C_{C2}(C_{C1} + C_L)} \tag{5.82}$$

Since C_{C1} is set by the required unity-gain bandwidth, and assuming G_{m1}, G_{m3} and C_L to be already set, (5.81) and (5.82) give the needed value of C_{C2}. Although it has been demonstrated here that ideal buffers provide a conceptually simple vehicle for the cancellation of the zeros, we will not stress these approaches any further because of the second-order effects of *real* buffers. In fact, the two approaches, as described above, prove to be inefficient especially in a low-voltage low-power context. Actually, the use of real voltage buffers unacceptably limits the output swing, while real current buffers –matching the requirement of very low input resistance– are expensive in terms of area and power consumption. Fortunately, we will show in 5.5.3 and 5.5.4 that both approaches can be simply modified so as to become suitable for practical applications.

For the sake of completeness, we shall first deal with the nulling resistor technique, which unfortunately is rather difficult to accomplish in a RNMC.

5.5.2 RHP Cancellation with Nulling Resistors

Figure 5.16 shows the compensation network including two nulling resistors, as customarily employed.

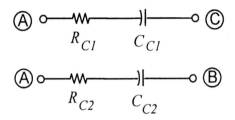

Fig. 5.16. Compensation network with two nulling resistors.

By using this network in the circuit in Fig. 5.7, the numerator of the open-loop gain in (5.78) becomes

$$N(s) = 1 + \left(R_{C1}C_{C1} + R_{C2}C_{C2} - \frac{C_{C2}}{G_{m2}} \right)s - C_{C1}C_{C2}\left(R_{C1}R_{C2} - \frac{1}{G_{m2}G_{m3}} - \frac{R_{C1}}{G_{m2}} \right)s^2$$

$$(5.83)$$

in which, as usual, only dominant terms are considered.

It can be shown that (5.83) provides real and negative zeros only with complex matching between R_{C1} and R_{C2} (by setting one of the two resistances equal to zero, it also is easy to verify that the RHP zero cannot be eliminated as the s^2 coefficient is always negative).

A more effective solution is that shown in Fig. 5.17, which uses only one resistor and leads to the following expression of $N(s)$

$$N(s) = 1 + \left[R_C(C_{C1} + C_{C2}) - \frac{C_{C2}}{G_{m2}} \right]s - \frac{C_{C1}C_{C2}}{G_{m3}}\left(\frac{1}{G_{m2}} - R_C \right)s^2 \quad (5.84)$$

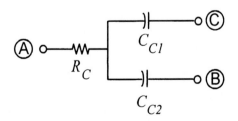

Fig. 5.17. Compensation network with one nulling resistor.

Setting $R_C = 1/G_{m2}$, (5.84) becomes

$$N(s) = 1 + \frac{C_{C1}}{G_{m2}} s \tag{5.85}$$

yielding only one negative zero. Of course, the denominator of the open-loop gain is still the same as in (5.78). In this case, it is convenient to have $G_{m2}=G_{m3}$. As we shall show, this choice allows a pole-zero cancellation to be achieved. Indeed, assuming that also

$$C_{C2}C_L > 4C_{C1}^{\ 2} \tag{5.86}$$

meaning that the determinant of the second order factor of (5.78) is positive, it follows that all poles are real and thus (5.78) becomes

$$A(s) = A_o \frac{1}{\left(1 + \dfrac{s}{p_1}\right)\left(1 + \dfrac{C_{C2}C_L}{G_{m3}C_{C1}} s\right)} \tag{5.87}$$

For a given phase margin we get

$$C_{C2} = \frac{C_{C1}^{\ 2} G_{m3}}{\tan(\phi) C_L G_{m1}} \tag{5.88}$$

Now, by substituting (5.88) in (5.86), condition (5.86) is satisfied if $G_{m2,3}/G_{m1} > 4\tan(\phi)$. Since a phase margin of about 60° is generally required, it follows that the transconductance of both the second and third stage must be at least seven times greater than the transconductance of the first stage. Since transconductances are usually set by other kinds of specifications, the application of this technique is implicitly limited.

5.5.3 RHP Cancellation with One Real Voltage Buffer

To achieve RHP cancellation, we can efficiently make use of only one voltage buffer in the inner loop, as shown in Figure 5.18.

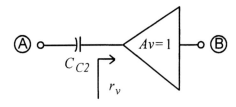

Fig. 5.18. Compensation network with one voltage follower.

By adopting this compensation network the output swing turns out to be completely preserved. In addition, we shall exploit the finite output resistance of the voltage follower to perform some simplifications as described below. Denoting this output resistance as r_v, the loop-gain transfer function is modified to

$$A(s) = \frac{A_o}{\left(1 + \dfrac{s}{p_1}\right)} \cdot \frac{1 + \left(r_v C_{C2} - \dfrac{C_{C1}}{G_{m2}G_{m3}r_v}\right)s - \dfrac{C_{C1}C_{C2}r_v}{G_{m2}G_{m3}R_2}s^2}{1 + \dfrac{C_{C2}\left(C_L + C_{C1} + G_{m3}r_vC_{C1}\right)}{G_{m3}C_{C1}}s + \dfrac{r_vC_{C2}C_L}{G_{m2}G_{m3}r_v}s^2} \approx$$

$$\approx \frac{A_o}{\left(1 + \dfrac{s}{p_1}\right)} \cdot \frac{\left(1 + r_v C_{C2}s\right)\left(1 - \dfrac{C_{C1}}{G_{m2}G_{m3}R_2}s\right)}{\left[1 + \dfrac{C_{C2}\left(C_L + C_{C1} + G_{m3}r_vC_{C1}\right)}{G_{m3}C_{C1}}s\right]\left[1 + \dfrac{C_{C1}C_Lr_v}{G_{m2}R_2\left(C_L + C_{C1} + G_{m3}r_vC_{C1}\right)}s\right]}$$

$$(5.89)$$

Relationship (5.89) includes one dominant LHP zero and a RHP zero that is now shifted to a very high frequency (since it is multiplied by the stage gain $G_{m2}R_2$). Moreover, there are two non-dominant poles which are real and negative under the condition (in practice usually met) $G_{m2}R_2C_{c2} > 2\, C_L$. These two poles are well approximated by the terms inside the square brackets in the second expression of (5.89). We can use the output resistance of the voltage follower to obtain some forms of simplification, and among the possible alternatives, we can set the value of this resistance equal to the transconductance of the last stage

$$r_v = \frac{1}{G_{m3}} \tag{5.90}$$

Substituting (5.90) in (5.89) the non-dominant poles and the two zeros result as

$$p_2 = \frac{C_{C1}}{(C_L + 2C_{C1})} \frac{G_{m3}}{C_{C2}} \tag{5.91}$$

$$p_3 = \frac{(C_L + 2C_{C1})}{C_L} \frac{G_{m3} G_{m2} R_2}{C_{C1}} \tag{5.92}$$

$$z_1 = \frac{G_{m3}}{C_{C2}} = \frac{C_L + 2C_{C1}}{C_{C1}} p_2 \tag{5.93}$$

$$z_2 = -\frac{G_{m3} G_{m2} R_2}{C_{C1}} \tag{5.94}$$

It is apparent that p_3 and z_2 are at a very high frequency (with $p_3 > |z_2|$) and their contribution to the phase margin can be neglected. Moreover $z_1 > 2p_2$ assures a monotonic behaviour for the loop gain module. The phase margin is then given by

$$\phi = \tan^{-1} \frac{p_2}{\omega_{GBW}} + \tan^{-1} \frac{\omega_{GBW}}{z_1} \tag{5.95}$$

from which we get

$$C_{C2} = C_{C1} \frac{G_{m3}}{G_{m1}} \frac{2\alpha}{(1-\alpha)\tan(\phi) + \sqrt{(1-\alpha)^2 \tan(\phi)^2 - 4\alpha}} \tag{5.96}$$

where parameter α is equal to

$$\alpha = \frac{C_{C1}}{C_L + 2C_{C1}} \tag{5.97}$$

Equation (5.96) can be simplified by observing that for the auspicious condition $C_{C1} < C_L$, parameter α is lower than $1/3$ and for practical phase margin values around $60°\text{-}70°$ we have

$$(1-\alpha)^2 \tan(\phi)^2 > 4\alpha \tag{5.98}$$

indicating that the zero can also be neglected when evaluating the phase margin. In conclusion, (5.96) can be approximated by (5.82).

5.5.4 RHP Cancellation with One Real Current Buffer

As can be deduced by returning to Fig. 5.14, the overall feedback current is given by the sum of the currents flowing into the two compensation capacitors. Thus, we need to use only *one* current buffer in the loop to break the forward path, as shown in Fig. 5.19, simplifying design and reducing power and area consumption. Since the overall feedback current is still the same, the loop-gain transfer function is again given by equation (5.80) and (5.82) still holds.

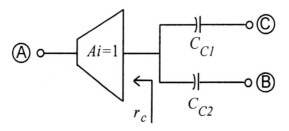

Fig. 5.19. Compensation network with one current follower.

If we consider the finite input resistance of the current follower, r_c, the loop gain will include another pole and two zeros, as shown below

$$A(s) = -A_0 \frac{1 + (C_{C1} + C_{C2})r_c s + \dfrac{C_{C1}C_{C2}r_c}{G_{m3}} s^2}{\left(1 + \dfrac{s}{p_1}\right)\left(1 + \dfrac{C_{C2}(C_{C1} + C_L)}{G_{m3}C_{C1}} s\right)\left[1 + \dfrac{C_L C_{C1} r_c}{(C_L + C_{C1})G_{m2}R_1} s\right]} \tag{5.99}$$

in which the two zeros have a negative real part. Besides, they are real if

$$r_c \le 4 \frac{C_{C1}C_{C2}}{G_{m3}(C_{C1} + C_{C2})^2} \tag{5.100}$$

that gives a higher limit for the current follower input resistance. If this condition is met, the expressions of the two zeros become

$$z_1 \approx \frac{1}{(C_{C1} + C_{C2})r_c} \tag{5.101}$$

$$z_2 \approx \frac{G_{m3}(C_{C1} + C_{C2})}{C_{C1}C_{C2}} \tag{5.102}$$

Choosing the highest value of r_c defined by equality in (5.100), it can be shown that z_2 is four times greater than z_1. Moreover, we have that $p_2 \ll p_3$ if $C_{C1}/C_{C2} \ll G_{m2}R_1/4$, which is a condition easily met in practice. Thus the second zero and the third pole in (5.99) are allocated well above the second pole, and do not appreciably modify the phase margin.

Finally, if $C_{C1} > C_{C2}$ we have that $p_2 < z_1$. This means that the first zero does not modify ω_{GBW}, but must be considered when evaluating the phase margin which is

$$\phi = \tan^{-1} \frac{G_{m3}}{G_{m1}} \frac{C_{C1}^2}{C_{C2}(C_{C1} + C_L)} + \tan^{-1} \frac{\omega_{GBW}}{z_1} \tag{5.103}$$

and is reduced to (5.82) if $z_1 \gg \omega_{GBW}$.

Thanks to the action of the negative zero, the approach adopting a current buffer is preferable to the one using a voltage buffer.

Chapter 6

FUNDAMENTAL FEEDBACK CONFIGURATIONS

In this chapter we will consider the four basic types of feedback amplifier: the *series-shunt*, the *shunt-series*, the *shunt-shunt* and the *series-series* configurations. These are used to realise voltage, current, transresistance and transconductance closed-loop amplifiers, respectively, and are capable of significantly reducing the dependence of forward transfer characteristics on the ill-controlled parameters implicit to the open-loop gain. Particularly, this chapter analyses first the low-frequency performance of these architectures, which are normally realised by multi-stage topologies, and subsequently gives frequency compensation guidelines. At this purpose, the results derived in Chapter 2 and Chapter 5 are extensively exploited.

6.1 SERIES-SHUNT AMPLIFIER

The AC schematic diagram of a *series-shunt* feedback amplifier is depicted in Fig. 6.1. In this circuit, the output voltage, v_o, is sampled by feedback network composed of the resistances R_E and R_F. The sampled voltage is fed back in such a way that the closed-loop input voltage, v_{Y1}, is the sum of v_{YX1} (the voltage across the input port of the amplifier) and v_{RE} (developed across R_E in the feedback subcircuit). Since $v_{Y1}=v_{YX1}+v_{RE}$, the output port of the feedback configuration can be viewed as connected in series with the amplifier input port. On the other hand, output voltage sampling constraints the net load current, i_{RL}, to the algebraic sum of the amplifier output port current, i_{Z2}, and the feedback network input current, i_{RF}. Accordingly, the output topology is indicative of a shunt connection of the feedback subcircuit to the amplifier output port. The fact that the voltage is fed back to a voltage-driven input port renders the driving point input resistance, r_{in}, of the closed-loop amplifier large, whereas the output resistance, r_{out}, is small. The resultant closed-loop amplifier is therefore best suited for voltage amplification, in the sense that the closed-loop voltage gain can be made approximately independent of source and load resistances. For large loop gain, this voltage loop gain is nominally determined by only the feedback subcircuit parameters.

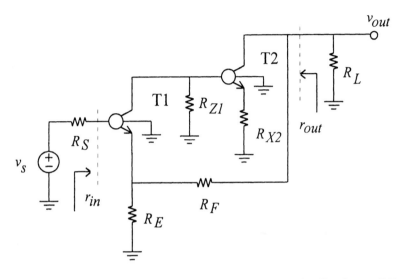

Fig. 6.1. AC schematic diagram of the series-shunt feedback amplifier.

The small signal model of the amplifier in Fig. 6.1 is shown in Fig. 6.2, where the model developed in section 2.4 and depicted in Fig. 2.7 has been used for the degenerated common X transistor T2.

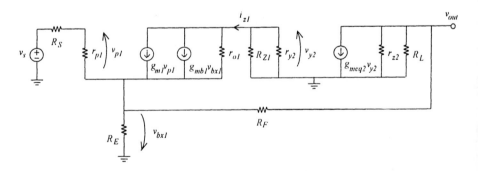

Fig. 6.2. Small-signal model of the series-shunt feedback amplifier.

As a practical rule, evaluation of the terms needed both in the Rosenstark and Choma methods or in the Blackman equations becomes simpler if one node of the controlled source is at ground potential. This means that for multi-transistor amplifiers we can profitably choose the controlled source among those associated to a common X transistor. According to this heuristic rule, in the small-signal model of the series-shunt amplifier we choose as controlled source P the transconductance g_{meq2}, and apply the Rosenstark method as described in the steps below. Observe that whenever possible we prefer evaluating the circuit parameters directly on the AC schematic. In our opinion, this is essential to develop the indispensable skills required to an analog circuit designer.

1) To evaluate the direct transmission term, G_o, we set $P=0$ ($g_{meq2}=0$). This, unless a load effect on terminal Z of T1, means eliminating transistor T2 and leading to the AC schematic diagram depicted in Fig. 6.3. It clearly represents a voltage follower whose transfer gain was derived in paragraph 2.6 and is expressed by relationships (2.28a) and (2.29). Then, substituting $R_E \| (R_L + R_F)$ and R_S to R_X and R_Y, respectively, and including term $R_L / (R_L + R_F)$ which takes into account the voltage partition at the output of the voltage buffer (resistance r_{z2} has been considered much higher than R_L), we get the gain, G_o, under the special condition of zero feedback

$$G_o = \frac{v_{out}}{v_s}\bigg|_{g_{meq2}=0} \approx \frac{R_L}{R_L + R_F}\frac{r_{y1}}{r_{y1} + R_S}\frac{g_{m1}R_E \| (R_L + R_F)}{(1+\lambda_{b1})g_{m1}R_E \| (R_L + R_F)+1} \quad (6.1)$$

Resistance r_{y1} is the resistance seen at the Y1 terminal of circuit in Fig. 6.3. In common cases, where $R_S < r_{y1}$ and the intrinsic voltage gain of the common Z configuration is close to the unity, relationship (6.1) can be further simplified into

$$G_o \approx \frac{R_L}{(1+\lambda_{b1})(R_L + R_F)} \quad (6.2)$$

which shows that this contribution is always lower than one. Thus, the direct transmission term, G_o, can be neglected without introducing appreciable errors in the evaluation of the closed-loop gain.

Under the same condition of controlled source set to zero, we can evaluate the driving point input and output resistances, $r_{in,ol}$, and $r_{out,ol}$, by using (2.24) and (2.25), respectively. We rearrange below the simplified expressions of $r_{in,ol}$, and $r_{out,ol}$ by considering $R_{z1}//r_{Y2} \ll r_{ol}$

$$r_{in,ol} = r_{p1} + (1+g_{m1}r_{p1})[R_E \| (R_L + R_F)] \quad (6.3)$$

$$r_{out,ol} = R_F + \frac{r_{p1} + R_S}{1 + (1 + \lambda_{b1})g_{m1}r_{p1}} \| R_E \approx R_F \qquad (6.4)$$

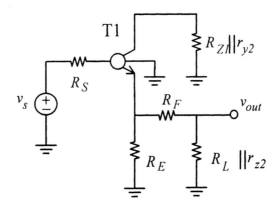

Fig. 6.3. AC schematic diagram of the series-shunt feedback amplifier setting $g_{meq2}=0$.

2) To evaluate the return ratio, we set v_s to zero and replace the original controlled current generator, $g_{meq2}v_{y2}$, with an independent current source, i. Again, transistor T2 can be eliminated while transistor T1 is in common Y configuration as illustrated in Fig. 6.4.

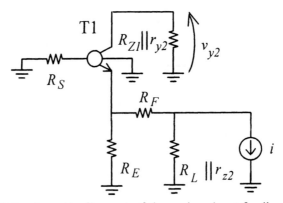

Fig. 6.4. AC schematic diagram of the series-shunt feedback amplifier setting $v_s = 0$.

The circuit can be simplified by considering the Norton equivalent of generator i and resistors R_E, R_F, and $R_L\|r_{z2}$, shown in Fig. 6.5

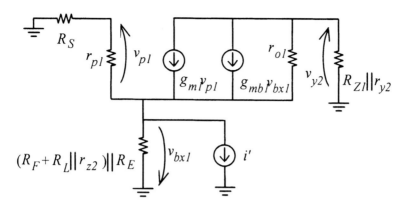

Fig. 6.5. Small-signal model of the series-shunt feedback amplifier setting $v_s = 0$.

where i' is given by

$$i' = \frac{R_L \| r_{z2}}{R_L \| r_{z2} + R_F} i \approx \frac{R_L}{R_L + R_F} i \tag{6.5}$$

and in which $R_L \ll r_{z2}$ is assumed in the approximation.

From (2.22), which expresses the current gain of a common Y configuration, assuming the input and output loss negligible (or in other words directly from (2.21)) we get (assuming $R_L \ll r_{z2}$)

$$\frac{i_{z1}}{i} \approx \frac{R_L}{R_L + R_F} \frac{(1+\lambda_{b1})g_{m1}r_{p1}}{(1+\lambda_{b1})g_{m1}r_{p1}+1} \frac{(R_F + R_L)\|R_E}{(R_F + R_L)\|R_E + r_{x1}} \approx \frac{R_L}{R_L + R_F} \tag{6.6}$$

Thus the return ratio, T, with respect to the critical parameter g_{meq2} is

$$T = -g_{meq2}\frac{v_{y2}}{i} = (R_{Z1}\|r_{y2})g_{meq2}\frac{i_{z1}}{i} \approx \frac{R_L}{R_L + R_F}(R_{Z1}\|r_{y2})g_{meq2} \tag{6.7}$$

3) Evaluate now the closed loop asymptotic gain, G_∞, by setting the parameter g_{meq2} infinitely large. By inspection of Fig. 6.2, since the current of the controlled current generator, $g_{meq2}v_{y2}$, is still finite, the voltage v_{y2} must be zero and this holds only if current i_{z1} is zero. The KCL at node Z1 implies

$$g_{m1}v_{p1} = \left(g_{mb1} + \frac{1}{r_{o1}}\right)v_{x1} \tag{6.8}$$

and the LKV at the input port gives

$$v_s = \left(\frac{R_S}{r_{p1}} + 1\right)v_{p1} + v_{x1} \tag{6.9}$$

Combining (6.8) with (6.9) we get

$$v_{x1} = \frac{v_s}{1 + \left(\dfrac{R_S}{r_{p1}} + 1\right)\left(\lambda_{b1} + \dfrac{1}{g_{m1}r_{o1}}\right)} \approx \frac{v_s}{1 + \lambda_{b1}} \tag{6.10}$$

in which the approximation presumes $R_S \ll r_p$. In conclusion, unless a small loss, the whole input voltage, v_s, is transferred across resistance R_E. According to Fig. 6.6, which derives from this consideration, we get

$$G_\infty = \frac{v_{out}}{v_s}\bigg|_{g_{meq2} \to \infty} = \left(1 + \frac{R_F}{R_E}\right)\frac{1}{1 + \lambda_{b1}} \tag{6.11}$$

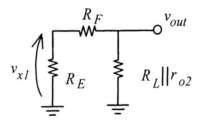

Fig. 6.6. Small-signal model which represents the virtual short-circuit condition.

Therefore, neglecting the contribution of G_o in the closed loop gain, the feedback factor, f, results

$$f = \frac{(1 + \lambda_{b1})R_E}{R_E + R_F} \tag{6.12}$$

The final closed loop gain expression is obtained by substituting G_o, T, and G_∞ into the Rosenstark relationship (3.7), which gives

$$G_F \approx \frac{\dfrac{1}{1+\lambda_{b1}}\left(1+\dfrac{R_F}{R_E}\right)\dfrac{R_L}{R_L+R_F}\left(R_{Z1}\|r_{y2}\right)g_{meq2}+\dfrac{R_L}{\left(R_L+R_F\right)\left(1+\lambda_{b1}\right)}}{1+\dfrac{R_L}{R_L+R_F}\left(R_{Z1}\|r_{y2}\right)g_{meq2}} \quad (6.13)$$

In the case we want to apply the Choma method, according to Section 3.4 we have to evaluate the null return ratio, T_R, instead of the asymptotic gain, G_∞. In particular, consider again the AC schematic diagram in Fig. 6.4 in which the input voltage source has not been nullified, as shown in Fig. 6.7a.

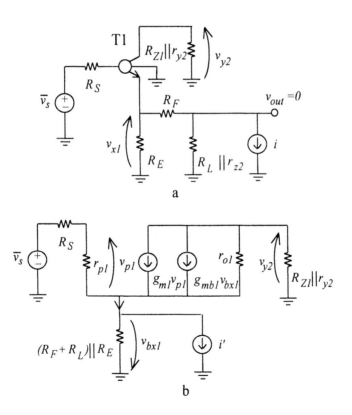

Fig. 6.7. Circuit for evaluating the null return ratio of series-shunt feedback amplifier: AC schematic diagram a), small-signal circuit b).

Since we must consider the output voltage to be zero, all the current, i, of the current generator (which replaces the critical controlled source) flows trough resistance R_F, and sets the voltage at node X1, v_{x1}, to be

$$v_{x1} = R_F i \tag{6.14}$$

By inspection of Fig. 6.7b we see that the current which sets voltage v_{y2} is equal to the sum of the currents which flow through resistances R_F and R_E, minus the current supplied by the input voltage source $\bar{v}_s$. Neglecting this last component, voltage v_{y2} is approximated by

$$v_{y2} \approx -\left(R_{Z1}\|r_{y2}\right)\left(1 + \frac{R_F}{R_E}\right)i \tag{6.15}$$

which means that the null return ratio is

$$T_R = -v_{y2} = \left(1 + \frac{R_F}{R_E}\right)\left(R_{Z1}\|r_{y2}\right)g_{m2} \tag{6.16}$$

Of course, a more accurate result can be achieved by using relationships reported in Section 2.5, but at the expense of simplicity and clearness. Moreover, (6.17) could be achieved through relationship (3.19), which relates the null return ratio with the parameters implicit in the Rosenstark method.

To evaluate the closed-loop input and output resistances we have firstly to calculate $T(0, R_L)$, $T(\infty, R_L)$, $T(R_S, 0)$ and $T(R_S, \infty)$ which are the return ratios under the specific conditions for the source resistance, R_S, and load resistance, R_L. The approximate expression of the return ratio derived in (6.7) does not include resistance R_S since we neglected the input loss. Therefore, (6.7) is representative of the condition $R_S = 0$.

$$T(0, R_L) = \frac{R_L}{R_L + R_F}\left(R_{Z1}\|r_{y2}\right)g_{meq2} \tag{6.17}$$

On the other hand, when R_S is equal to infinite, in the equivalent model in Fig. 6.5 we must nullify the current flowing through R_S. Thus, to evaluate $T(\infty, R_L)$ we can use the circuit in Fig. 6.8 with $i_s = 0$ and being

$$i' = \frac{R_L\|r_{Z2}}{R_L\|r_{Z2} + R_F}i \approx \frac{R_L}{R_L + R_F}i \tag{6.18}$$

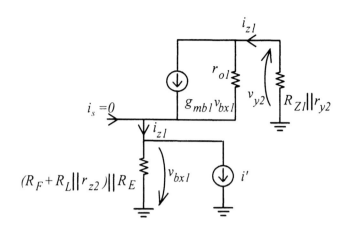

Fig. 6.8. Small-signal model for the evaluation of the Null Return Ratio and $T(\infty, R_L)$ in the series-shunt feedback amplifier.

Neglecting the effect of current $g_{mb1}v_{bx1}$, current gain i_{z1}/i is

$$\frac{i_{z1}}{i} \approx \frac{R_L}{R_L + R_F} \frac{R_E\|(R_L + R_F)}{R_E\|(R_L + R_F) + r_{o1} + R_{Z1}\|r_{y2}}$$ (6.19)

and hence we get

$$T(\infty, R_L) = -g_{meq2}\frac{v_{y2}}{i} = g_{meq2}\left(R_{Z1}\|r_{y2}\right)\frac{i_{z1}}{i} \approx$$

(6.20)

$$\approx g_{meq2}\frac{R_L}{R_L + R_F}\frac{R_E(R_F + R_L)R_{Z1}\|r_{y2}}{(R_E + R_F + R_L)r_{o1}}$$

After substituting (6.3), (6.17) and (6.20) into (3.23) we find the input resistance. However, since $T(\infty,R_L)$ is generally lower than 1, r_{in} can be approximated to

$$r_{in} \approx r_{in,ol}\left[1 + \frac{R_L}{R_L + R_F}\left(R_{Z1}\|r_{y2}\right)g_{meq2}\right] \approx$$

(6.21)

$$\approx \frac{R_L R_E}{R_L + R_F + R_E}g_{m1}r_{p1}\left(R_{Z1}\|r_{y2}\right)g_{meq2}$$

It is apparent that for MOS transistor implementations r_{in} is infinite since $r_{in,ol}$ is infinite. Parameters $T(R_S, 0)$ and $T(R_S, \infty)$ can be directly evaluated from (6.7) and are given by

$$T(R_S,0) = 0 \tag{6.22}$$

$$T(R_S,\infty) = \left(R_{z1}\|r_{y2}\right)g_{meq2} \tag{6.23}$$

Thus, according to (3.24) the output resistance is

$$r_{out} = \frac{r_{out,ol}}{1+\left(R_{Z1}\|r_{y2}\right)g_{meq2}} \approx \frac{R_F}{\left(R_{Z1}\|r_{y2}\right)g_{meq2}} \tag{6.24}$$

and, as expected, compared to its open-loop value is reduced by approximately the loop gain.

6.1.1 Series-shunt Amplifier with Buffer

When practical design constraints lead to a feedback resistance, R_F, too much higher than load resistance, R_L, the loop gain could become too low, and we get a proportional reduction of the feedback benefits. This problem can be circumvented by inserting a voltage follower between the output port of transistor T2 and the node to which the load termination and the input terminal of the feedback subcircuit are incident. The resultant circuit diagram, is shown in Fig. 6.9.

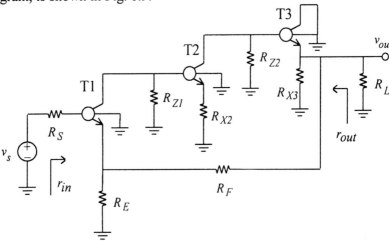

Fig. 6.9. AC schematic diagram of the series-shunt feedback amplifier with output buffer.

Transistor T3, highly reduces the open-loop output resistance, $r_{out,ol}$, which becomes quite independent of feedback resistance, R_F. The foregoing improvement can be confirmed through an analysis of the small-signal model of the modified amplifier in Fig. 6.9. To this end, assume transistor T3 a quite ideal voltage follower with unitary gain and with an input resistance much higher than resistance R_{Z2}. The loop gain of the circuit in Fig. 6.9 can be evaluated by using the small signal model in Fig 6.10, which can be further simplified by using a Norton equivalent current generator. After some approximations the ultimate circuit model is that shown in Fig. 6.11.

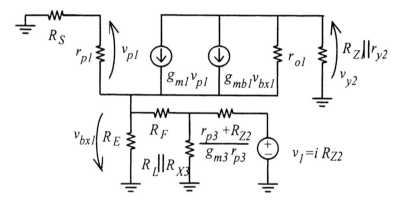

Fig. 6.10. Small-signal model to evaluate the loop gain of the circuit in Fig. 6.9.

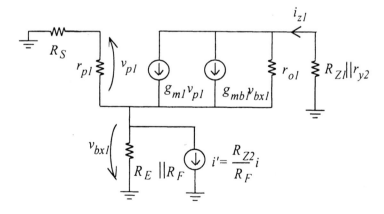

Fig. 6.11. Approximated equivalent model of circuit in Fig. 6.10.

Consequently, (6.6) can now be rewritten as

$$\frac{i_{z1}}{i} = \frac{(1 + \lambda_{b1})g_{m1}r_{p1}}{(1 + \lambda_{b1})g_{m1}r_{p1} + 1} \frac{R_{Z2}}{R_F} \frac{R_F \| R_E}{R_F \| R_E + r_{x1}} \tag{6.25}$$

and, the return ratio with respect to the critical parameter g_{meq2}, is

$$T = (R_{Z1} \| r_{y2})g_{meq2} \frac{i_{z1}}{i} \approx \frac{R_{Z2}}{R_F}(R_{Z1} \| r_{y2})g_{meq2} \tag{6.26}$$

The fundamental difference with the previous case is that instead of factor $R_L/(R_L + R_F)$, which introduces open loop gain reduction, we have now factor R_{Z2}/R_F, which can be simply made greater than one. Moreover, asymptotic gain, G_∞, is unchanged, while forward gain, G_o, is further reduced. Indeed, resistance R_L in (6.2) is now replaced by the parallel between R_L itself and the resistance seen at node X of transistor T3 (i.e.,

$\dfrac{r_{p3} + R_{z2}}{1 + (1 + \lambda_{b3})g_{m3}r_{p3}} \| R_{X3} \| R_L$). It is apparent that since the circuit now exhibits

a higher open loop gain and a lower forward gain, the closed loop gain, G_F, tends to be more close to G_∞. Of course, the improved topology provides also advantages in terms of input resistance, increased by the greater loop gain

$$r_{in} \approx (R_E \| R_F)g_{m1}r_{p1} \frac{R_{Z2}}{R_F}(R_{Z1} \| r_{y2})g_{meq2} \tag{6.27}$$

and in terms of output resistance, which is reduced both for the greater loop gain and for the lower open loop output impedance provided by voltage buffer T3

$$r_{out} \approx \frac{R_F}{R_{z2}} \frac{r_{p3} + R_{Z2}}{(R_{Z1} \| r_{y2})g_{meq2}r_{p3}g_{m3}} \tag{6.28}$$

6.2 SHUNT-SERIES AMPLIFIER

While the series-shunt feedback circuit well behaves as a voltage amplifier, the shunt-series configuration, whose AC schematic diagram is depicted in Fig. 6.12, is best suited for implementing a current amplifier. In the subject circuit, the current on node X of transistor T2, equal to the output signal current, i_{out}, (approximately equal for BJT's), is sampled by the feedback network formed by resistors R_E and R_F. The sampled current is fed back to a current-driven input port. Thus, the resultant driving point output resistance is large, and the driving point input resistance is small. These characteristics allow for a closed loop current gain, $G_F(R_S, R_L) = i_o/i_S$, that is relatively independent of source and load resistances as well as insensitive to transistor parameters.

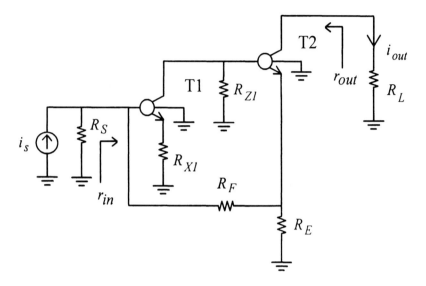

Fig. 6.12. AC schematic diagram of the shunt-series feedback amplifier.

To analyse the circuit in Fig. 6.12, consider its small-signal model illustrated in Fig. 6.13, where the model of the common X configuration with a degenerative resistance (see Fig. 2.4) has been used for transistor T2, and assume the transconductance g_{meq1} as critical parameter P.

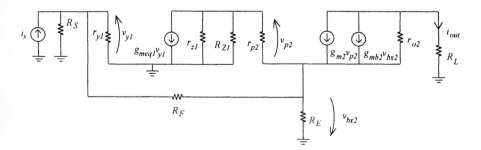

Fig. 6.13. Small-signal model of the shunt-series feedback amplifier.

1) To evaluate the direct transmission term, G_o, we set $P=0$ ($g_{meq1}=0$). This means switching off transistor T1, then taking into account the load effects of T1 we can consider the circuit AC schematic diagram depicted in Fig. 6.14, and its small-signal model in Fig. 6.15

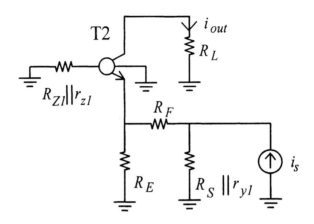

Fig. 6.14. AC schematic diagram of the shunt-series feedback amplifier setting $g_{meq1}=0$.

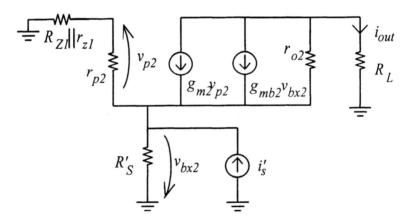

Fig. 6.15. Small-signal model of the shunt-series feedback amplifier setting $g_{meq1}=0$.

Resistance R'_S and current i'_S in Fig. 6.15 are given by

$$R'_S = \left(R_F + R_S \| r_{y1}\right) \| R_E \tag{6.29}$$

$$i_s' = \frac{R_S \| r_{y1}}{R_F + R_S \| r_{y1}} i_s \tag{6.30}$$

Since circuit in Fig. 6.15 represents a common Y configuration, it can be analysed by using its equivalent model in Fig. 6.16, in which A_{yi2} and r_{z2} are defined in (2.21) and (2.11).

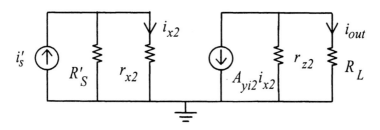

Fig. 6.16. Equivalent model of the circuit in Fig. 6.15.

Then using relationships developed in paragraph 2.5 we obtain

$$G_o = \frac{i_{out}}{i_s}\bigg|_{g_{meq1}=0} = \frac{R_S'}{R_S' + r_{x2}} \frac{(1 + \lambda_{b2})g_{m2}r_{p2}}{(1 + \lambda_{b2})g_{m2}r_{p2} + 1} \frac{r_{z2}}{R_L + r_{z2}} \tag{6.31}$$

which is a quantity always lower than one. Moreover, we can easily evaluate the input and output resistance, $r_{in,ol}$, and $r_{out,ol}$, by using the results in (2.20a) and (2.11), respectively. Note that to evaluate the input resistance we have to consider the circuit scheme in Fig. 6.13

$$r_{in,ol} \approx \left[R_F + \frac{r_{p2} + R_{Z1}}{1 + (1 + \lambda_{b2})g_{m2}r_{p2}} \| R_E \right] \| r_{y1} \tag{6.32}$$

$$r_{out,ol} = r_{z2} \approx g_{m2}(1 + \lambda_{b2})r_{o2}[(r_{p2} + R_{Z1} \| r_{z1}) \| R_E] \tag{6.33}$$

2) To evaluate the return ratio we set i_s to zero and replace the original controlled current generator, $g_{meq1}v_{y1}$, with an independent one, i. Now, as can be deduced from the equivalent circuit shown in Fig. 6.17, transistor T2 works as a voltage follower, and voltage v_{y1} is a portion of the voltage at node X of T2.

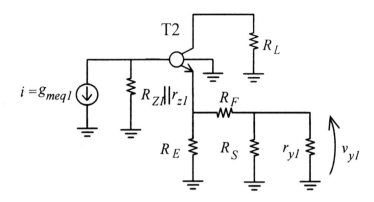

Fig. 6.17. AC schematic diagram of the shunt-series feedback amplifier
setting $i_s = 0$.

To simplify calculation, assume that r_{o2} is much higher than R_L and R_E,
and that R_{Z1} is the lower resistance at node Y of transistor T2. Using
equation (2.38c), expressing the voltage gain of a common Z transistor, the
loop gain results to be

$$T = -v_{y1} \approx \frac{g_{m2}R_E\|(R_F + R_S\|r_{y1})}{(1 + \lambda_{b2})g_{m2}R_E\|(R_F + R_S\|r_{y1}) + 1} \frac{R_S\|r_{y1}}{R_S\|r_{y1} + R_F} g_{meq1}R_{Z1} \quad (6.34)$$

which, neglecting the voltage loss between terminals Y and X of T2, lastly
simplifies to

$$T \approx \frac{R_S\|r_{y1}}{R_S\|r_{y1} + R_F} g_{meq1}R_{Z1} \qquad (6.35)$$

3) Now we evaluate the closed loop asymptotic gain. By inspection of
Fig. 6.13, we realise that setting the parameter g_{meq1} infinitely large leads v_{y1}
to equal zero, which in turn means that all the input current, i_s, flows into
feedback resistance, R_F. In addition, since a finite value of current $g_{meq1}v_{y1}$
would cause v_{y1} to be different from zero, term $g_{meq1}v_{y1}$ itself must equal
zero. With these considerations, we can model the circuit with that in Fig.
6.18.

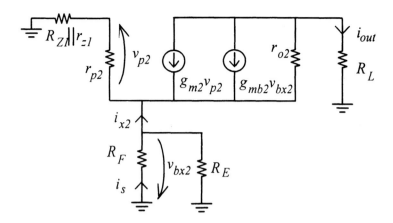

Fig. 6.18. Small-signal circuit to evaluate the asymptotic gain of the shunt-series feedback amplifier.

By inspection of fig. 6.18 we easily find that the current entering into terminal X of transistor T2 is equal to

$$i_{x2} = \left(1 + \frac{R_F}{R_E}\right) i_s \tag{6.36}$$

hence, neglecting resistance r_{o2}, we get

$$G_\infty = \left. \frac{i_{out}}{i_s} \right|_{g_{meq1} \to \infty} = \frac{(1 + \lambda_{b2}) g_{m2} r_{p2}}{1 + (1 + \lambda_{b2}) g_{m2} r_{p2}} \left(1 + \frac{R_F}{R_E}\right) \tag{6.37}$$

Therefore, substituting (6.32), (6.35) and (6.37) into the Rosenstark relationship, exact expression of the closed loop gain of shunt-series feedback amplifier can be found. For practical cases where the loop gain is much greater than one, the closed loop gain is very well approximated by the asymptotic one.

Finally, we have to calculate the resultant input and output resistances through terms $T(0, R_L)$, $T(\infty, R_L)$, $T(R_S, 0)$ and $T(R_S, \infty)$. From (6.35) we can simply calculate $T(0, R_L)$ and $T(\infty, R_L)$, whose expressions are

$$T(0, R_L) = 0 \tag{6.38}$$

$$T(\infty, R_L) = \frac{g_{m2}r_{p2}}{(1+\lambda_{b2})g_{m2}r_{p2}+1+\dfrac{r_{p2}}{R_E\|(R_F+r_{y1})}} \frac{r_{y1}}{r_{y1}+R_F} g_{meq1}R_{Z1} \qquad (6.39)$$

Moreover, the expression of $T(R_S, 0)$ is exactly as given by (6.35). Indeed this equation was calculated assuming R_L negligible compared to r_{o2}. However, in the opposite case of infinitely large R_L we have to evaluate under this hypothesis the voltage gain of the common Z in Fig. 6.17 and the general result in (2.29) becomes[1]

$$A_{zv}\big|_{R_z\to\infty} = \frac{R_X}{R_X+r_p} \approx \frac{R_E\|(R_F+R_S)}{R_E\|(R_F+R_S)+r_{p2}} \qquad (6.40)$$

Thus, term $T(R_S, \infty)$ results to be

$$T(R_S,\infty) \approx \frac{R_E\|(R_F+R_S)}{R_E\|(R_F+R_S)+r_{p2}} \frac{R_S}{R_S+R_F} g_{meq1}R_{Z1} \qquad (6.41)$$

Now we are able to apply Blackman equations to obtain the closed-loop input and output resistance expressions. The input resistance is reduced by the specific loop gain in (6.39)

$$r_{in} = \frac{R_F}{\dfrac{g_{m2}r_{p2}g_{meq1}R_{Z1}}{(1+\lambda_{b2})g_{m2}r_{p2}+1+\dfrac{r_{p2}}{R_E}}} \approx \frac{1+\lambda_{b2}}{g_{meq1}R_{Z1}} R_F \qquad (6.42)$$

The output resistance although reduced by term (6.41) is also heavily increased by the loop gain (6.35)

---

[1] From the exact equation (2.28a) assuming $R_Z = R_L\to\infty$, we get $A_{zv} = \dfrac{r_p\|R_X}{r_p}$ and neglecting the input and output voltage loss in (2.29) we obtain $A_v \approx A_{zv}$

$$r_{out} = r_{out,ol} \frac{1 + \dfrac{g_{m2}r_{p2}}{(1+\lambda_{b2})g_{m2}r_{p2} + 1 + \dfrac{r_{p2}}{R_E\|(R_F+R_S)}} \dfrac{R_S}{R_S+R_F} g_{meq1} R_{Z1}}{1 + \dfrac{R_E\|(R_F+R_S)}{R_E\|(R_F+R_S)+r_{p2}} \dfrac{R_S}{R_S+R_F} g_{meq1} R_{Z1}} \qquad (6.43)$$

thus, as clearly indicated by the approximated relation (6.44) we globally get the expected increase in the output resistance

$$r_{out} \approx g_{m2}r_{o2}\left(r_{p2}\|R_E\right)\left(1 + \frac{R_{Z1}+r_{p2}}{R_E\|(R_F+R_S)}\right) \qquad (6.44)$$

6.3 SHUNT-SHUNT AMPLIFIER

The AC schematic diagram of the third type of single loop feedback amplifier, the shunt-shunt amplifier, is drawn in Fig. 6.19. A cascade interconnection of three transistors, T1, T2 and T3, forms the open loop, while the feedback subcircuit comprises a single resistance, R_F. The output voltage, v_{out}, is sampled by R_F and converted into a current which is fed back to the input port. Therefore, both the driving point input and output resistance are very small. Accordingly, the circuit operates best as a transresistance amplifier in that its closed loop transresistance, $G_F(R_S, R_L) = v_{out}/i_s$, is nominally invariant with source resistance, load resistance and transistor parameters.

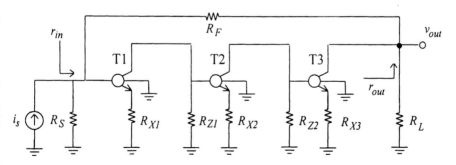

Fig. 6.19. AC schematic diagram of the shunt-shunt feedback amplifier.

Considering the equivalent small-signal model of shunt-shunt circuit shown in Fig. 6.20, we can arbitrarily choose transconductance g_{meq1} as parameter P (but other different choices do not lead to any fundamental difference).

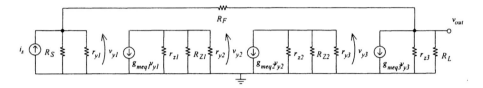

Fig. 6.20.Small-signal model of the shunt-shunt feedback amplifier.

1) We set $P = 0$ ($g_{meq1} = 0$) then taking into account the load effects (input resistance of T1 and output resistance of T3), the circuit small-signal model depicted in Fig. 6.21 is considered. The feedforward transresistance and the corresponding input and output resistances are

$$G_o = \frac{v_{out}}{i_s}\bigg|_{g_{meq1}=0} = \frac{R_S\|r_{y1}}{R_S\|r_{y1} + R_F + R_L\|r_{z3}} R_L\|r_{z3} \qquad (6.45)$$

$$r_{in,ol} = \left(R_F + R_L\|r_{z3}\right)\|r_{y1} \qquad (6.46)$$

$$r_{out,ol} = \left(R_F + R_S\|r_{y1}\right)\|r_{z3} \qquad (6.47)$$

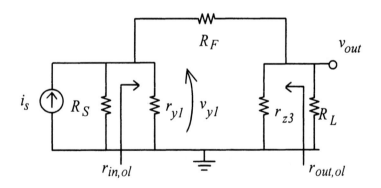

Fig. 6.21. Small-signal model of the shunt-shunt feedback amplifier setting $g_{meq1} = 0$.

2) Set i_s to zero and replace the controlled current generator with an independent current source, i. Transistor T1 can be considered switched off, and the resulting AC model is that in Fig. 6.22, which leads to the small-signal model in Fig. 6.23. The return can be immediately evaluated from Fig. 6.23, or even directly from Fig. 6.22, and is given by

$$T = g_{meq1}\left(R_{Z1}\|r_{y2}\right)g_{meq2}\left(R_{Z2}\|r_{y3}\right)g_{meq3}\frac{R_L\|r_{z3}}{R_L\|r_{z3} + R_F + R_S\|r_{y1}}R_S\|r_{y1} \quad (6.48)$$

where $r_{z1} \gg R_{Z1}$ and $r_{z2} \gg R_{Z2}$ was assumed for simplicity

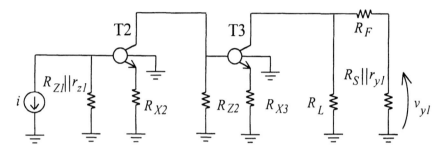

Fig. 6.22. AC schematic diagram of the shunt-shunt feedback amplifier setting $i_s = 0$.

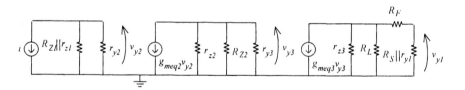

Fig. 6.23. Small-signal model of the shunt-shunt feedback amplifier setting $i_s = 0$.

3) Set the parameter the parameter g_{meq1} infinitely large. By inspection of Fig. 6.20, since the voltage v_{y2} is still finite, the voltage v_{y1} must be zero. This means that all the input current flows through feedback resistance R_F, thus setting the output voltage v_{out} to $-R_F i_s$. The asymptotic transresistance results to be

$$G_\infty = \left.\frac{v_o}{i_s}\right|_{g_{meq1}\to\infty} = -R_F \quad (6.49)$$

Hence, substituting G_o, T and G_∞ into the Rosenstark relationship (3.7), we get the closed loop transresistance G_F, that, in general, due to the very high loop gain is almost coincident with the asymptotic gain

$$G_F \approx -R_F \tag{6.50}$$

To calculate the closed-loop input and output resistances we can derive $T(0, R_L)$, $T(\infty, R_L)$, $T(R_S, 0)$ and $T(R_S, \infty)$ directly from (6.48). Thus we have

$$T(0, R_L) = 0 \tag{6.51}$$

$$T(\infty, R_L) = g_{meq1}\left(R_{Z1}\|r_{y2}\right)g_{meq2}\left(R_{Z2}\|r_{y3}\right)g_{meq3}\frac{R_L\|r_{z3}}{R_L\|r_{z3} + R_F + r_{y1}}r_{y1} \tag{6.52}$$

$$T(R_S, 0) = 0 \tag{6.53}$$

$$T(R_S, \infty) = g_{meq1}\left(R_{Z1}\|r_{y2}\right)g_{meq2}\left(R_{Z2}\|r_{y3}\right)g_{meq3}\frac{r_{z3}}{r_{z3} + R_F + R_S\|r_{y1}}R_S\|r_{y1} \tag{6.54}$$

and the input and output resistances are

$$r_{in} = \frac{\left(R_F + R_L\|r_{z3}\right)\|r_{y1}}{g_{meq1}\left(R_{Z1}\|r_{y2}\right)g_{meq2}\left(R_{Z2}\|r_{y3}\right)g_{meq3}\left(R_L\|r_{z3}\right)\dfrac{r_{y1}}{R_L\|r_{z3} + R_F + r_{y1}}} \tag{6.55}$$

$$r_{out} = \frac{\left(R_F + R_S\|r_{y1}\right)\|r_{z3}}{g_{meq1}\left(R_{Z1}\|r_{y2}\right)g_{meq2}\left(R_{Z2}\|r_{y3}\right)g_{meq3}\left(R_S\|r_{y1}\right)\dfrac{r_{z3}}{R_S\|r_{y1} + R_F + r_{z3}}} \tag{6.56}$$

6.4 SERIES-SERIES AMPLIFIER

Fig. 6.24 depicts the AC schematic of the series-series feedback amplifier. Three transistors, T1, T2, and T3, are embedded in the open loop amplifier and the feedback subcircuit is provided by resistor R_F combined with local feedback resistors R_{E1} and R_{E2}. The feedback network samples the output current, i_{out}, and compares it with the input voltage. Therefore, both the driving point input and output resistance are very high, and, accordingly, the circuit operates best as a transconductance amplifier in that its closed loop transresistance, $G_F(R_S, R_L) = i_{out}/v_s$, is nominally invariant with source resistance, load resistance and transistor parameters.

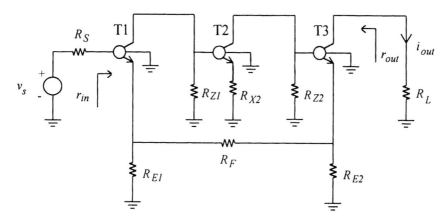

Fig. 6.24. AC schematic diagram of the series-series feedback amplifier.

From a comparison of series-series feedback circuit in Fig. 6.24 with series-shunt feedback amplifier with output buffer in Fig. 6.9, it is apparent that the only difference between the two amplifiers is the load resistance at node Z of transistor T3. Therefore, unless some minor differences we can follow the same analysis we have previously developed for the series-shunt amplifier with output buffer. In particular choose as controlled source P the transconductance g_{meq2}, and proceed with the following steps.

1) Set $P=0$ ($g_{meq2}=0$). Assuming transistor T1 implementing an ideal voltage buffer and transistor T3 implementing an ideal current buffer, the feedforward transconductance can be directly derived by inspection of Fig. 6.24, and results to be

$$G_o = \left. \frac{i_{out}}{v_s} \right|_{g_{meq2}=0} \approx -\frac{1}{R_F} \tag{6.57}$$

Moreover, the open-loop input and output resistances are about equal to

$$r_{in,ol} \approx r_{p1} + \left(1 + g_{m1}r_{p1}\right)\left(R_{E1} \| R_F\right) \tag{6.58}$$

$$r_{out,ol} \approx r_{o3} + \left[1 + g_{m3}\left(1 + \lambda_{b3}\right)r_{o3}\right]r_{p3} \| \left(R_{E2} + R_F\right) \tag{6.59}$$

2) Replace the controlled current generator with an independent current source, i, and set the input voltage source to zero. In this case, neglecting the effect of load resistance on the accuracy of the voltage follower implemented by transistor T3, we get the same return ratio as the series-shunt feedback amplifier with output buffer in Fig. 6.9. Hence, it is about equal to (6.27)

$$T \approx \frac{R_{Z2}}{R_F} R_{Z1} \| r_{y2} g_{meq2} \tag{6.60}$$

Consider the transconductance g_{meq2} infinitely large. Since the voltage v_{y3} is still finite, the voltage v_{y2} must be zero. This implies that the current provided by terminal Z of transistor T1 must be zero, and hence the voltage v_{y1} and the associated current given by the voltage source must be zero. In conclusion, the asymptotic transconductance can be evaluated on the equivalent AC model in Fig. 6.25.

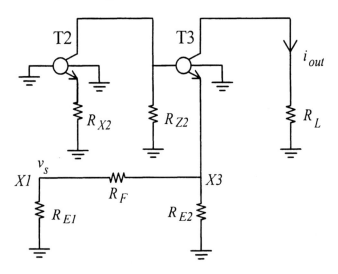

Fig. 6.25. AC schematic diagram to evaluate the asymptotic gain of the series-series feedback amplifier.

In particular, the voltage at terminal X of transistor T1 (node X1 in Fig. 6.25) is equal to the source voltage. Furthermore, transistor T1 does not supply current. The current through resistance R_{E1}, (i.e., v_s / R_{E1}) is thus equal to that flowing through resistance R_F. Besides, the voltage on node X3, which defines the current of resistance R_{E2}, is

$$v_{x3} = \left(1 + \frac{R_F}{R_{E1}} \right) v_s \tag{6.61a}$$

and the current which flows into node X3 is equal to

$$i_{x3} = -\left(\frac{v_s}{R_{E1}} + \frac{v_{x3}}{R_{E2}}\right) = -\left[\frac{1}{R_{E1}} + \frac{1}{R_{E2}}\left(1 + \frac{R_F}{R_{E1}}\right)\right]v_s \qquad (6.61b)$$

which leads to

$$G_\infty = \frac{i_{out}}{v_s}\bigg|_{g_{meq2}\to\infty} \approx -\left(\frac{1}{R_{E1}} + \frac{1}{R_{E2}} + \frac{R_F}{R_{E1}R_{E2}}\right) \qquad (6.62)$$

The closed-loop transconductance is given combining (6.57), (6.60) and (6.62) in (3.7), and since the loop gain is generally very large, the closed loop transconductance is quite equal to the asymptotic one G_∞.

The assumption of ideal voltage and current buffer for T1 and T3, respectively, in the evaluation of the return ratio, means that we are neglecting the load effect of resistance R_S and R_L. Hence, the return ratios calculate nullifying the source or the load resistance are almost equal to the value given by relationship (6.60)

$$T(R_S,0) \cong T(0,R_L) \cong \frac{R_{z2}}{R_F} R_{Z1}\|r_{y2}g_{meq2} \qquad (6.63)$$

Besides, we have to evaluate the other two specific return ratios $T(\infty, R_L)$ and $T(R_S, \infty)$. To do this, return to Fig. 6.24. To calculate $T(\infty, R_L)$ assume again that transistor T3 is an ideal voltage buffer. Thus voltage v_{y3}, which is equal to $-g_{meq2}R_{Z2}$, appears unchanged at terminal X3, and the voltage at node Y2 can be found by considering two voltage partitions in X1 and Y2 respectively

$$v_{y2} \cong -\frac{(r_{o1} + R_{Z1}\|r_{y2})\|R_{E1}}{R_F + (r_{o1} + R_{Z1}\|r_{y2})\|R_{E1}} \frac{R_{Z1}\|r_{y2}}{r_{o1} + R_{Z1}\|r_{y2}} g_{meq2}R_{Z2} \cong \qquad (6.64)$$

Thus $T(\infty, R_L)$ is

$$T(\infty, R_L) \approx \frac{R_{E1}\|R_F}{R_{E1}\|R_F + r_{o1} + R_{Z1}\|r_{y2}} \frac{R_{Z1}\|r_{y2}}{R_F} g_{meq2}R_{Z2} \cong$$

$$(6.65)$$

$$\approx \frac{R_{E1}}{R_{E1} + R_F} \frac{R_{Z1}\|r_{y2}}{r_{o1}} g_{meq2}R_{Z2}$$

which is usually lower than 1.

To calculate $T(R_S, \infty)$ consider transistor T1 working as an ideal current buffer having an input resistance at terminal X to be ideally zero. Thus, current g_{meq2} is partitioned, and the part which flows through resistance R_F is found also at terminal Z of T1. After some calculation we get

$$T(R_S, \infty) \approx \left(R_{Z1}\|r_{y2}\right)g_{meq2} \frac{R_{Z2}}{R_{Z2} + r_{p3} + \left(R_F\|R_{E2}\right)} \frac{R_{E2}}{R_{E2} + R_F} \qquad (6.66)$$

In conclusion, combining (6.64), (6.65) and (6.66) in the Blackman relationships we get

$$r_{in} \approx r_{in,ol}\left(1 + g_{meq2}R_{Z1}\|r_{y2}\frac{R_{Z2}}{R_F}\right) \frac{\left(R_{E1} + R_F\right)r_{ol}}{\left(R_{E1} + R_F\right)r_{ol} + g_{meq2}R_{Z1}\|r_{y2}R_{Z2}R_{E1}} \qquad (6.67)$$

$$r_{out} \approx r_{out,ol}\left(1 + g_{meq2}R_{Z1}\|r_{y2}\frac{R_{Z2}}{R_F}\right) \frac{1}{1 + \dfrac{g_{meq2}R_{Z1}\|r_{y2}R_{Z2}}{R_{Z2} + r_{p3} + \left(R_F\|R_{E2}\right)}\dfrac{R_{E2}}{R_{E2} + R_F}} \qquad (6.68)$$

6.5 A GENERAL VIEW OF SINGLE-LOOP AMPLIFIERS

From the cases discussed in the previous paragraphs of this chapter, it clearly turns out that the Rosenstark method, and in particular the use of the asymptotic gain, leads to a simplified description of a feedback amplifier in terms of an ideal one with infinite loop gain. As the reader should well know, a particular form of this approximation is common practice in opamp analysis and design. This conducts to the well-known principle of virtual short circuit. This approximation is of course adequately verified in feedback circuits with large loop-gain, and allows to a calculation of the circuit closed-loop gain that is considerably simpler then in the original circuit and particularly suited for pencil-and-paper computation.

As a conclusion of this part of the chapter, we want to show that the four classes of feedback amplifiers above analysed can be ultimately represented with a circuit in which the transistor network is substituted by an ideal differential amplifier having infinite voltage gain, infinite input resistance, zero output resistance, and eventually an output current-controlled current source.

From this point of view, the *ideal* series-shunt feedback amplifier is illustrated in Fig. 6.26. It is implemented by the differential amplifier in non-inverting configuration. Its transfer function is coincident with the asymptotic term previously obtained in (6.11) given by

$$\frac{v_{out}}{v_s} = 1 + \frac{R_F}{R_E} \qquad (6.69)$$

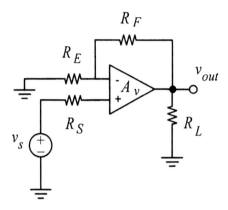

Fig. 6.26. Ideal scheme of a series-shunt feedback amplifier.

The *ideal* shunt-series feedback amplifier is illustrated in Fig. 6.27. It is implemented with and ideal differential amplifier and a current-controlled current source with a gain, a_i. All the input current goes through the feedback resistance, R_F, and sets the differential amplifier output voltage.

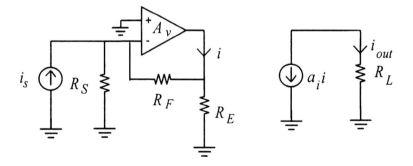

Fig. 6.27. Ideal scheme of a shunt-series feedback amplifier.

The output current, i, provided by the differential amplifier is sensed and delivered through the current-controlled generator at the load resistance. The complete transfer function is

$$\frac{i_{out}}{i_s} = a_i\left(1 + \frac{R_F}{R_E}\right) \tag{6.70}$$

and is equivalent to (6.37).

The *ideal* shunt-shunt feedback amplifier is depicted in Fig. 6.28. It performs the current to voltage conversion thanks to the virtual ground exhibited by the differential amplifier. Its transfer function is

$$\frac{v_{out}}{i_s} = -R_F \tag{6.71}$$

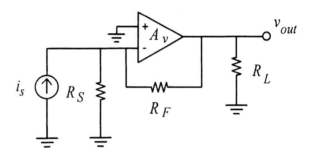

Fig. 6.28. Ideal scheme of a shunt-shunt feedback amplifier.

The *ideal* series-series feedback amplifier is shown in Fig. 6.29. It performs a voltage to current conversion. Indeed, due to the virtual short circuit, the input voltage appears at the inverting node of the differential amplifier, thus generating a current through resistance, R_{E1}, which is also equal to that through resistance, R_F. The differential amplifier output voltage is then set by v_s and the current through R_F. This, divided by R_{E2} sets the differential amplifier output current, i, that is replicated at the output of the series-series feedback amplifier through the current-controlled crrent source of gain a_i. The ideal transfer function of the series-series feedback amplifier is hence given by

$$\frac{i_{out}}{v_s} = -a_i\left[\frac{1}{R_{E2}}\left(1 + \frac{R_F}{R_{E1}}\right) + \frac{1}{R_{E1}}\right] \tag{6.72}$$

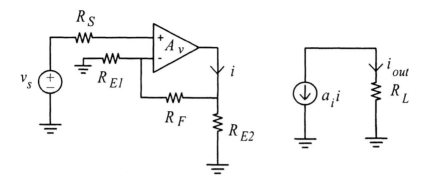

Fig. 6.29. Ideal scheme of a series-series feedback amplifier.

It is apparent that by using the virtual short circuit principle, transfer functions of the ideal feedback amplifiers can straightforwardly be evaluated without the need of feedback theory. Of course, finite gain and finite resistances of the differential amplifier can be used to better model a real feedback amplifier, but the use of more accurate models can lead to a loss in simplicity of the circuit analysis. Sometimes, the analysis of a network containing even only one such a "real" differential amplifier is so difficult that is more efficient to apply the Rosenstark method.

The ideal feedback amplifiers presented in this paragraph have the purpose of giving more insight into the feedback amplifiers performance. Their use allows both to improve the intuitive perception of the four basic feedback circuit topology as well as their inherent performance and to quickly achieve approximate input-output transfer functions.

6.6 FREQUENCY COMPENSATION OF THE FUNDAMENTAL CONFIGURATIONS

As discussed in the two previous chapters, an amplifier operated in feedback configuration requires compensation. To be more precise, we have to guarantee an adequate phase margin within the *specific portion* of the circuit that, closed in feedback, provides the system loop gain. This key point means that, to provide stability, we have to consider *soley* the circuit path utilised to evaluate the return ratio and subject it to the compensation techniques described in Chapter 5.

In the following, we shall present simple guidelines for the compensation of the fundamental configurations which are amenable to pencil-and-paper evaluation. To this end, we will reduce multi-pole systems into two-pole ones. Of course, this constitutes a rough simplification that, nevertheless, affords a deep understanding of the most significant elements associated wit2h the compensation step. More accurate results involving second-order effects that are required to further refine the compensation process are achieved by using circuit simulators like SPICE [VNP80].

6.6.1 Frequency Compensation of the Series-Shunt Amplifier

The schematic shown in Fig. 6.30 is the high-frequency small-signal equivalent circuit of the series-shunt amplifier in Fig. 6.1. In the scheme, the load capacitance C_L includes the output capacitance, C_{o2}, of transistor T2.

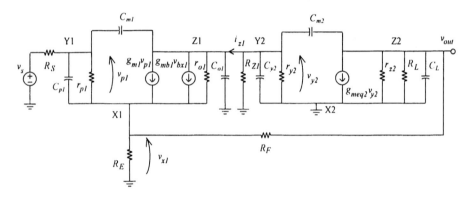

Fig. 6.30. High-frequency small-signal model of the series-shunt feedback amplifier.

To evaluate the return ratio we must set voltage source v_s to zero. By initial circuit inspection we find that the loop gain is made up of the gain-stage provided by transistor T2, closed in loop through the feedback resistors R_F and R_E, and transistor T1. In particular, transistor T1 is in common Y configuration and works as a current buffer.

Since inside the loop we have only one inverting gain stage with both input and output high-resistance nodes, to compensate the circuit we can profitably exploit the Miller approach by connecting the compensation capacitor, C_C, across terminals Y and Z of transistor T2.

This generates a dominant pole at node Y2. The second pole is due to the output capacitance, and we neglect the other high-frequency poles within the loop. Now, to perform compensation we have to simply follow the procedure developed in Chapter 5, on the equivalent circuit in Fig. 6.31, also equivalent to the one in Fig. 5.2 except for the additional capacitor C_1. Note however, that since C_C is a large capacitor, the parasitic capacitor C_1 can be neglected.

Of course, to avoid positive zeros we must use one of the techniques discussed in the same chapter.

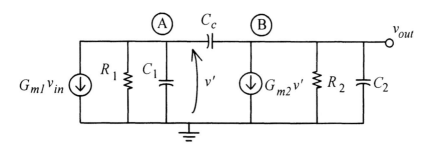

Fig. 6.31. Equivalent small-signal model used for compensation.

By comparing the circuit in Fig. 6.31 and the one in Fig. 6.30, we find that the transconductance of the second stage coincides with transconductance g_{meq2}

$$G_{m2} = g_{meq2} \tag{6.73}$$

Moreover, the second-stage output resistance accounts for the feedback resistors according to

$$R_2 = R_L \| r_{z2} \| (R_F + R_E \| r_{x1}) \tag{6.74}$$

where r_{x1} is evaluated according to (2.20f) and is a low-valued resistance. Thus, (6.74) generally simplifies to

$$R_2 \approx R_L \| r_{z2} \| R_F \tag{6.75}$$

The equivalent transconductance G_{m1} can be evaluated by applying a test voltage, v_{z2}, at node Z of transistor T2, and evaluating the short circuit current, i_{sc}, at node Z of transistor T1. Since transistor T1 is in a common Y configuration, the current through terminal Z1 is approximately equal to that flowing into terminal X1. Consequently, transconductance G_{m1} is found by analysing the circuit model in Fig. 6.32, which implies

$$G_{m1} = \frac{1}{r_{x1}} \frac{r_{x1} \| R_E}{r_{x1} \| R_E + R_F} \approx \frac{1}{R_F} \tag{6.76}$$

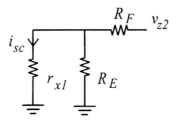

Fig. 6.32. Equivalent small-signal model to evaluate G_{m1}.

Resistance R_1 is that found at node Z1-Y2, and is equal to

$$R_1 = R_{Z1} \| r_{z1} \| r_{y2} \tag{6.77}$$

where r_{z1} is evaluated by short-circuiting node Z of transistor T2. Hence from (2.11) it is equal to

$$r_{z1} = r_{o1} + \left[1 + g_{m1} \left(\frac{r_{p1}}{r_{p1} + R_S} + \lambda_{b1} \right) r_{o1} \right] (r_{p1} + R_S) \| R_E \| R_F \tag{6.78}$$

which is usually a large resistance and can be neglected in (6.77).

Finally, we evaluate capacitors C_1 and C_2. The first capacitance is equal to $C_{o1} + C_{y2}$, but, as already mentioned, it is redundant because of the dominant contribution given by the compensation capacitor. The second capacitance is equal to the load capacitance (usually much greater than the intrinsic capacitances). This is an important contribution as it determines the second pole. Once all the circuit parameters in Fig. 6.31 have been identified, the compensation steps straightforwardly follow those given in Chapter 5.

The compensation of a series-shunt feedback amplifier with an output buffer is achieved in a substantially similar manner.

Observe that the approach described above is implicitly equivalent to the one customarily used by circuit designers, that requires breaking the feedback loop and suitably updating the impedance levels. This approach is pictorially described in Fig. 6.33. It is also worth mentioning that analytical methods like the one described in [T92] have been developed to find the loop gain without breaking the loop. These techniques are particularly efficient if associated with circuit simulation programs.

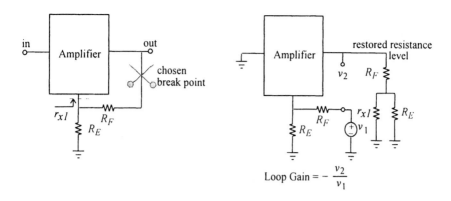

Fig. 6.33. Evaluating the loop gain by breaking the loop.

6.6.2 Frequency Compensation of the Shunt-Series Amplifier

The schematic shown in Fig. 6.34 is the high-frequency small-signal circuit of the series-shunt amplifier already depicted in Fig. 6.12. Note that load capacitance C_L includes the output capacitance, C_{o2}, of transistor T2.

For the shunt-series amplifier, the circuit path for evaluating the return ratio comprises the gain stage (provided by transistor T1) closed in unitary loop by transistor T2 (which is in common Z configuration and works as a voltage buffer) and the feedback resistors. It is worth noting that the load impedance is outside the loop, and, hence, does not play any role in the compensation process.

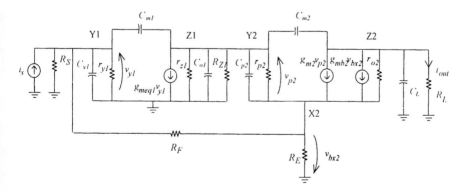

Fig. 6.34. High-Frequency small-signal model of the shunt-series feedback amplifier.

Again, like the series-shunt amplifier, we have only one inverting gain stage with high-resistance input and output nodes, which correspond to the Y and Z nodes, respectively, of transistor T1. Thus Miller compensation can be usefully applied (although dominant-pole compensation could in principle be performed at node Z1 or even at Y1).

By using the equivalent circuit in Fig. 6.31, the transconductance of the second stage G_{m2} now coincides with the equivalent transconductance of transistor T1

$$G_{m2} = g_{meq1} \tag{6.80}$$

and the second-stage output resistance is

$$R_2 = R_{Z1} \| r_{z1} \| r_{y2} \tag{6.81}$$

where r_{y2} is the input resistance (at terminal Y2) of the common Z transistor. It can be calculated using the expressions in Chapter 2, however, it is large and can be generally neglected.

The transconductance of the first stage, G_{m1}, is evaluated by applying a test voltage source at node Y2, calculating the short-circuit current at node Y1, and then taking their ratio. If we approximate the voltage gain between Y2 and X2 to be exactly unitary, we get

$$G_{m1} = \frac{1}{R_F} \tag{6.82}$$

Resistance R_1 is the equivalent one at node Y of transistor T1, and is equal to

$$R_1 \approx R_S \| r_{y1} \| R_F \tag{6.83}$$

where node X of transistor T2 is assumed to have an output resistance small enough to be considered as a ground connection.

Finally, we calculate the equivalent capacitor C_2 which is given by the capacitive contribution at node Z1-Y2

$$C_2 = C_{o1} + C_{y2,\text{TOT}} \tag{6.84}$$

in which $C_{y2,TOT}$ can be evaluated by suitably modelling transistor T2 according to the method described in section 2.7.2 and using equation (2.46), which yield

$$C_{y2,TOT} = C_{y2} + C_{m2}\left(1 + g_{meq2}R_L\right) \tag{6.85}$$

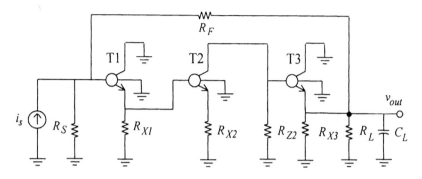

Fig. 6.35. AC schematic diagram of a buffered shunt-shunt feedback amplifier.

6.6.3 Frequency Compensation of the Shunt- Shunt Amplifier

To analyse the compensation of the shunt-shunt amplifier, for simplicity, we will refer to the circuit in Fig. 6.35 instead of using the one in Fig. 6.19. It exhibits only one amplifying stage (transistor T2) within two voltage buffering stages provided by transistor T1 and T3. Thus, the equivalent resistance at the gain stage input is low (that of terminal X1) and Miller compensation becomes impractical. Instead, we have to use, dominant-pole compensation at node Y3 which is the only high-impedance node. The equivalent resistance which, associated with the compensation capacitor, sets the dominant pole is

$$R_2 = R_{Z2}\|r_{z2}\|r_{y3} \approx R_{Z2} \tag{6.86}$$

The second pole arises at the output terminal, as is associated with the usually large load capacitance.

It is worth noting that the original scheme in Fig. 6.19 was made up of three (inverting) gain stages. This structure usually includes three low-frequency poles (two at the high-resistance internal nodes, Y2 and Y3, and one at the output due to the high load capacitance). In this case, the most suitable compensation technique is a variant of the nested-Miller one and is called hybrid nested Miller. We shall not discuss this technique here, for details interested readers are referred to [EH95].

6.6.4 Frequency Compensation of the Series-Series Amplifier

Drawing the high-frequency small-signal circuit of the series-series amplifier in Fig. 6.24 is left to the reader.

For the series-series amplifier, the circuit path for evaluating the return ratio includes the gain stage provided by transistor T2, transistor T3 (which acts as a voltage buffer), the feedback resistors, and transistor T1 which acts as a current follower. Again, the load impedance is outside the loop, and hence does not play any role in the compensation process. Given the presence of an inverting gain stage with both input/output high-resistance terminals, Miller compensation is the most suitable technique.

Thus, transconductance G_{m2} of the equivalent model in Fig. 6.31 must be assumed to be the equivalent transconductance g_{meq2} of transistor T2

$$G_{m2} = g_{meq2} \tag{6.87}$$

and resistors R_3 and R_1 are given by

$$R_2 = R_{Z2} \| r_{z2} \| r_{y3} \approx R_{Z2} \tag{6.89}$$

$$R_1 \approx R_{Z1} \tag{6.90}$$

In (6.90) the loading effect of the feedback network can be neglected because of the buffering operation of transistor T1.

Finally, assuming T1 and T3 to be ideal current and voltage follower, respectively, we get

$$G_{m1} \approx \frac{1}{R_F} \tag{6.91}$$

Chapter 7

HARMONIC DISTORTION

Semiconductor devices are inherently nonlinear. For example, Bipolar transistors in forward active region exhibit an exponential relationship between the collector current and the base-emitter voltage, while in saturated MOS transistors the drain current approximately depends on the square of the gate-source voltage. Therefore, circuits made up with transistors or, more generally, with real active components exhibit a certain amount of nonlinearity, and this means that the relationship between their input and the output variables is not so ideally linear as assumed in the previous chapters. Usually, active devices used for analog signal processing applications are operated in a quasi-linear region. Thus the linearity assumption is almost verified especially when signals with small amplitude are processed. However, designers are asked to evaluate the limits of the linear approximation or to characterise the effects of nonlinear distortion in circuits and systems used as linear blocks [S99][1]. To achieve these targets harmonic distortion analysis is customary employed.

Consider the open loop amplifier A_{NL} in Fig. 7.1 with its DC *nonlinear* transfer characteristic $x_o=A_{NL}(x_i)$. When nonlinearities are small, that is the transcharacteristic is characterised by gradual slope changes, the circuit is said to operate under *low-distortion conditions*[2]. This implies, in other words, that transistors do not leave the active region, and small-signal analysis can be used to produce meaningful results. Harmonic distortion in this case is usually calculated with the series expansion of the nonlinear DC

---

[1] Linear distortion arises in a linear amplifier which has a non constant frequency response in the frequency domain [S99].

[2] For a rigorous definition of the low-distortion condition see [OS93].

transfer characteristic. Let us assume that it is well represented by the first three power terms

$$x_o = A_{NL}(x_i) \approx a_1 x_i + a_2 x_i^2 + a_3 x_i^3 \tag{7.1}$$

Assuming now the incremental input voltage be a pure sinusoidal tone $x_i = X_i \cos(\omega t)$, the output signal becomes[3]

$$x_o = b_0 + b_1 \cos(\omega t) + b_2 \cos(2\omega t) + b_3 \cos(3\omega t) \tag{7.2}$$

where terms b_i are

$$b_0 = \frac{a_2}{2} X_i^2 \tag{7.3}$$

$$b_1 = a_1 X_i + \frac{3}{4} a_3 X_i^3 \tag{7.4}$$

$$b_2 = \frac{a_2}{2} X_i^2 \tag{7.5}$$

$$b_3 = \frac{a_3}{4} X_i^3 \tag{7.6}$$

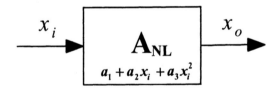

Fig. 7.1. Nonlinear open-loop amplifier.

Due to the amplifier nonlinearity, the ideal sinusoid at the input changes its shape at the output. Indeed, the output signal is a superposition of a constant term, represented by the coefficient b_0, a sinusoidal waveform with a frequency equal to that at the input multiplied by the coefficient b_1

---

[3] Remember that $\cos^2 x = \dfrac{1 + \cos 2x}{2}$ and $\cos^3 x = \dfrac{3\cos x + \cos 3x}{4}$.

(*fundamental* component), and two other sinusoidal waveforms having a frequency twice and three times greater than that of the input signal, multiplied by the coefficients b_2 and b_3, respectively (*second* and *third harmonic* components). To outline the weight of the harmonics, the harmonic distortion factors are defined as given below [S70]

$$HD_2 = \frac{|b_2|}{|b_1|} \approx \frac{1}{2} \frac{a_2}{a_1} X_i \qquad (7.7a)$$

$$HD_3 = \frac{|b_3|}{|b_1|} \approx \frac{1}{4} \frac{a_3}{a_1} X_i^2 \qquad (7.7b)$$

where the gain compression [MW95], which arises in term b_1 and is due to coefficient a_3, have been neglected. It is worth noting that the harmonic factors increase with the input amplitude.

In order to allow a simple comparison with the closed-loop cases that will be developed in the following paragraphs, the harmonic distortion factors can be also referred to the amplitude of the output fundamental component, $X_o \approx a_1 X_i$

$$HD_2 = \frac{1}{2} \frac{a_2}{a_1^2} X_o \qquad (7.8a)$$

$$HD_3 = \frac{1}{4} \frac{a_3}{a_1^3} X_o^2 \qquad (7.8b)$$

Of course, the two above equations can be used to compare the linearity performance of two different amplifiers but at the same (fundamental) output signal level.

Alternatively, we can represent the input signal by the expression $x_i = X_i e^{j\omega t}$ and the output signal, through (7.1), becomes $x_o = c_1 e^{j\omega t} + c_2 e^{j2\omega t} + c_3 e^{j3\omega t} = a_1 X_i e^{j\omega t} + a_2 X_i^2 e^{j2\omega t} + a_3 X_i^3 e^{j3\omega t}$. Thus, to obtain the same distortion factors as in (4) we have to define $HD_2 = \frac{1}{2} \frac{|c_2|}{|c_1|}$

and $HD_3 = \frac{1}{4} \frac{|c_3|}{|c_1|}$. As we will show this representation is useful to characterise nonlinear systems in the frequency domain.

7.1 HARMONIC DISTORTION AT LOW FREQUENCY

In this section we shall analyse the influence of feedback on harmonic distortion for low-frequency input signals. In other words, we consider the input signal frequency lower than the cut-off frequency of the loop gain, which can be therefore assumed constant, i.e. $T(j\omega) = T_o$.

7.1.1 Nonlinear Amplifier with Linear Feedback

The classical theory of feedback amplifiers asserts that negative feedback reduces harmonic distortion by the loop-gain [GM93], [LS94]. Let us consider the same amplifier in Fig. 7.1 characterised by the same nonlinear function given in (7.1), and feed a fraction f of the output signal back to the input, as shown in Fig. 7.2. This means to close the amplifier in loop with a linear feedback, f, and obtaining a return ratio T_o equal to fa_1. It is well known that the harmonic distortion terms given by (7.7a) and (7.7b) are reduced by the factors $(1+T_o)^2$ and $(1+T_o)^3$, respectively. Alternatively, we obtain a reduction by a factor $(1+T_o)$ on the harmonic distortion factors referred to the output signal magnitude.

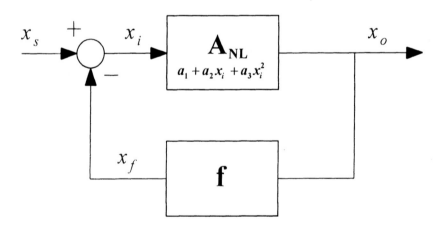

Fig. 7.2. Nonlinear amplifier with linear feedback.

Following the approach described in [PM91], a more accurate result of the harmonic distortion factors of a closed-loop amplifier can be obtained. Indeed, for the feedback amplifier in Fig. 7.2 we have

$$x_i = x_s - x_f = x_s - fx_o \tag{7.9}$$

hence relationship (7.1) can be rewritten as

$$x_o = a_1(x_s - fx_o) + a_2(x_s - fx_o)^2 + a_3(x_s - fx_o)^3 \tag{7.10}$$

The output signal can be expressed as a new power series with the source signal as independent variable

$$x_o \approx a_1' x_i + a_2' x_i^2 + a_3' x_i^3 \tag{7.11}$$

where coefficients a_1', a_2', and a_3' can be obtained by interpreting the power series as a Taylor's series

$$x_o = \left. \frac{dx_o}{dx_s} \right|_{0,0} x_s + \frac{1}{2} \left. \frac{d^2 x_o}{dx_s^2} \right|_{0,0} x_s^2 + \frac{1}{6} \left. \frac{d^3 x_o}{dx_s^3} \right|_{0,0} x_s^3 \tag{7.12}$$

Taking the derivatives of (7.10) and considering that $x_o = 0$ when $x_s = 0$, we obtain

$$a_1' = \frac{a_1}{1 + T_o} \tag{7.13}$$

$$a_2' = \frac{a_2}{(1 + T_o)^3} \tag{7.14}$$

$$a_3' = \frac{a_3(1 + fa_1) - 2a_2^2 f}{(1 + fa_1)^5} \tag{7.15}$$

that through relationships (7.7) and (7.8) lead to

$$HD_{2fl} = \frac{1}{2} \frac{a_2}{a_1} \frac{1}{(1 + T_o)^2} X_s = \frac{1}{2} \frac{a_2}{a_1^2} \frac{1}{1 + T_o} X_o \tag{7.16}$$

$$HD_{3fl} = \frac{1}{4}\frac{a_3}{a_1}\frac{1 - \dfrac{2fa_2^2}{a_3(1+T_o)}}{(1+T_o)^3}X_s^2 = \frac{1}{4}\frac{a_3}{a_1^3}\frac{1 - \dfrac{2fa_2^2}{a_3(1+T_o)}}{1+T_o}X_o^2 \tag{7.17}$$

in which subscript "*fl*" stands for *linear feedback*.

By inspection of (7.17) we see that for amplifiers where coefficient a_3 is negligible, the third harmonic is still determined by a_2. Moreover, the third harmonic distortion can be minimised if

$$\frac{f}{1+T_o} = \frac{a_3}{2a_2^2} \tag{7.18}$$

and for $T_o \gg 1$ (7.17) and (7.18) simplify to

$$HD_{3fl} = \frac{1}{4}\frac{a_3}{a_1}\frac{1 - \dfrac{2a_2^2}{a_1 a_3}}{(1+T_o)^3}X_s^2 = \frac{1}{4}\frac{a_3}{a_1^3}\frac{1 - \dfrac{2a_2^2}{a_1 a_3}}{1+T_o}X_o^2 \tag{7.19a}$$

$$\frac{2a_2^2}{a_1 a_3} = 1 \tag{7.19b}$$

7.1.2 Nonlinear Amplifier with Nonlinear Feedback

When also the feedback network is made up of active components (for instance, when MOS transistors working in triode region are employed as feedback elements instead of pure linear resistances, [IF94]), the feedback network cannot be considered ideally linear as previously done. Evaluation of the distortion of a feedback amplifier where both the amplifier and the feedback network introduce substantial nonlinearities was carried out in [PP981], and is developed in the following.

First, consider the nonlinear behaviour of the feedback path according to

$$x_f = F_{NL}(x_{out}) = f_1 x_{out} + f_2 x_{out}^2 + f_3 x_{out}^3 \tag{7.20}$$

The input signal, x_s, can be written as

$$x_s = x_i + x_f \tag{7.21}$$

hence, after substituting (7.1) into (7.20), and again substituting the resulting equation into (7.21), we get

$$x_s = \left(1 + f_1 a_1\right)x_i + \left(f_1 a_2 + f_2 a_1^2\right)x_i^2 + \left(f_1 a_3 + 2 f_2 a_1 a_2 + f_3 a_1^3\right)x_i^3 \qquad (7.22)$$

We can invert a nonlinear function, $x_s(x_i)$, represented by a power series up to the third-order term

$$x_s(x_i) = c_1 x_i + c_2 x_i^2 + c_3 x_i^3 \qquad (7.23)$$

into

$$x_i(x_s) = K_1 x_s + K_2 x_s^2 + K_3 x_s^3 \qquad (7.24)$$

by using the conversion formulas [KO91], [WM95] reported below

$$K_1 = \frac{1}{c_1} \qquad (7.25)$$

$$K_2 = -\frac{c_2}{c_1^3} \qquad (7.26)$$

$$K_3 = \frac{1}{c_1^3}\left[-\frac{c_3}{c_1} + 2\left(\frac{c_2}{c_1}\right)^2\right] \qquad (7.27)$$

Therefore, x_i is given by

$$x_i = \frac{1}{1 + f_1 a_1} x_s - \frac{f_1 a_2 + f_2 a_1^2}{\left(1 + f_1 a_1\right)^3} x_s^2$$

$$(7.28)$$

$$- \frac{1}{\left(1 + f_1 a_1\right)^3}\left[\frac{f_1 a_3 + 2 f_2 a_1 a_2 + f_3 a_1^3}{1 + f_1 a_1} - 2\left(\frac{f_1 a_2 + f_2 a_1^2}{1 + f_1 a_1}\right)^2\right]x_s^3$$

and combining (7.28) with (7.1), taking only the first three power terms, we get

$$x_o = a_1 K_1 x_s + \left(a_1 K_2 + a_2 K_1^2\right) x_s^2 + \left(a_1 K_3 + 2a_2 K_1 K_2 + a_3 K_1^3\right) x_s^3 \qquad (7.29)$$

In conclusion, being $T_o = a_1 f_1$, the second and third harmonic distortion factors are

$$HD_{2f} = \frac{1}{2}\left(\frac{K_2}{K_1} + \frac{a_2}{a_1} K_1\right) X_s = \frac{1}{2}\frac{a_2 - f_2 a_1^3}{a_1 \left(1 + T_o\right)^2} X_s = \frac{1}{2}\frac{a_2 - f_2 a_1^3}{a_1^2 \left(1 + T_o\right)} X_o \quad (7.30)$$

$$HD_{3f} = \frac{1}{4}\left(\frac{K_3}{K_1} + 2\frac{a_2}{a_1} K_2 + \frac{a_3}{a_1} K_1^2\right) X_s^2 =$$

$$= \frac{1}{4}\frac{a_3}{a_1}\frac{1 - f_3 \dfrac{a_1^4}{a_3} - \dfrac{2 f_1 a_2^2 + 4 f_2 a_1^2 a_2 - 2 f_2^2 a_1^5}{a_3 \left(1 + T_o\right)}}{\left(1 + T_o\right)^3} X_s^2 = \qquad (7.31)$$

$$= \frac{1}{4}\frac{a_3}{a_1^3}\frac{1 - f_3 \dfrac{a_1^4}{a_3} - \dfrac{2 f_1 a_2^2 + 4 f_2 a_1^2 a_2 - 2 f_2^2 a_1^5}{a_3 \left(1 + T_o\right)}}{1 + T_o} X_o^2$$

A more compact form of the second and third harmonic distortion coefficients can be obtained considering that the return ratio, T_o, is usually much greater than one. Hence, after normalising the amplifier terms a_2, a_3 to the amplifier gain a_1, and the feedback terms f_2, f_3 to the feedback linear term, f_1, (by defining $a_{2N} = a_2/a_1$, $a_{3N} = a_3/a_1$, $f_{2N} = f_2/f_1$ and $f_{3N} = f_3/f_1$) we get

$$HD_{2f} = \frac{1}{2}\left(\frac{1}{Ta_1} a_{2N} - f_{2N}\right) X_o \qquad (7.32)$$

$$HD_{3f} = \frac{1}{4}\left[\frac{1}{Ta_1^2}\left(a_{3N} - 2a_{2N}^2\right) - \left(f_{3N} - 2f_{2N}^2\right) - 4\frac{1}{Ta_1} a_{2N} f_{2N}\right] X_o^2 \quad (7.33)$$

By inspection of (7.32) and (7.33), it is apparent that feedback does not reduce the nonlinearity of the feedback network. Thus, we cannot obtain an amplifier having a nonlinearity lower than that of the feedback network, and even small nonlinearity terms of the feedback networks cannot be neglected, but they must be taken into account.

In order to evaluate the different weight between the nonlinearity of the amplifier and that of the feedback network, it is useful to write the two coefficients when the amplifier is linear (i.e., with $a_2 = 0$ and $a_3 = 0$)

$$HD_{2fn} = -\frac{1}{2}\frac{f_2 a_1^2}{(1+T_o)^2} X_s = -\frac{1}{2}\frac{f_2 a_1}{(1+T_o)} X_o \qquad (7.34)$$

$$HD_{3fn} = -\frac{1}{4}\frac{f_3 a_1^3 - \dfrac{2 f_2^2 a_1^4}{(1+T_o)}}{(1+T_o)^3} X_s^2 = -\frac{1}{4}\frac{f_3 a_1 - \dfrac{2 f_2^2 a_1^2}{(1+T_o)}}{1+T_o} X_o^2 \qquad (7.35)$$

As expected, comparison of (7.34) and (7.35) with (7.16) and (7.17), which refer to the case of nonlinear amplifier with linear feedback, shows that the feedback path is more critical than the forward path. Indeed, assuming the nonlinearity for both the amplifier and the feedback network to be equal, which means $a_2 = f_2$ and $a_3 = f_3$, for the same output magnitude, relationship (7.16) is lower than (7.34) by a factor a_1^2, and relationship (7.17) is lower than (7.35) by about a_1^4. Moreover, it is worth noting that for negative feedback, distortion due to the feedback network has an opposite sign to that due to the amplifier.

A more compact and clear representation of the harmonic distortion in a nonlinear amplifier with nonlinear feedback is

$$HD_{2f} = HD_{2fl} + HD_{2fn} \qquad (7.36a)$$

$$HD_{3f} = HD_{3fl} + HD_{3fn} + 4 HD_{2fl} HD_{2fn} \qquad (7.36b)$$

In conclusion, the second and third harmonic distortion terms can be compactly represented by relationships (7.36) which are only a simple function of the second and third harmonic distortion of the whole feedback network evaluated in two particular cases:
- a nonlinear amplifier with linearised feedback network
- a linearised amplifier with nonlinear feedback network.

This consideration can be particularly interesting from a design point of view, since other than allowing us to get more insight into the circuit behaviour and its final performance, permits to evaluate all the harmonic distortion factors through separate calculation (or simulation) of the two couples of terms HD_{2fl}, HD_{3fl} and HD_{2fn}, HD_{3fn} [PP981].

7.2 HARMONIC DISTORTION IN THE FREQUENCY DOMAIN

In the previous paragraphs, both the amplifier and the feedback network were assumed to be frequency independent. This hypothesis is clearly a rough approximation. Transistors have parasitic capacitances which cause the gain and even the nonlinear amplifier coefficients to vary with frequency. Yet, high-gain feedback circuits must be frequency compensated to ensure closed-loop stability, while the feedback network can include reactive (usually capacitive) components. Therefore, the previous expressions can be used with reasonable accuracy only under the hypothesis of low-frequency input signals.

In general, evaluation of harmonic distortion of a *dynamic* system requires complex calculation involving Volterra series or even Wiener series [BR71], [MSE72], [NP73], [WG99]. Nevertheless, under the assumption of low-distortion conditions –which means in practice, that the amplifier output is not saturated and transistors do not leave their active region of operation– we can use the usual small-signal analysis to produce accurate results. Let us start our discussion by considering amplifiers in open-loop configuration.

7.2.1 Open-loop Amplifiers

To render the analysis sufficiently general, we will refer to two-stage amplifiers, that adequately model real amplifiers (the obtained results could then be extended also to multi-stage topologies, as well). Besides, we simplify analysis by separating the effect of nonlinearities of the first and second stage. These two cases are illustrated in Fig. 7.3a and 7.3b. Of course, in real amplifiers both the two phenomena coexist as nonlinearity can contemporarily come from the input and the output sections. Nevertheless, this simplification is instructive and even representative of actual cases. Indeed, the first scheme (Fig. 7.3a) exemplifies a conventional op-amp or a CMOS OTA with a nonlinear output stage. In this event the output section operates in large-signal conditions and its nonlinear behaviour is hence exacerbated. The second scheme (Fig. 7.3b) seems uncommon. Later, we will demonstrate that this case models the high-frequency distortion in single-stage amplifiers. Besides, it can exemplify amplifiers operated under large common-mode input signals, responsible for the generation of nonlinearities in the input stage.

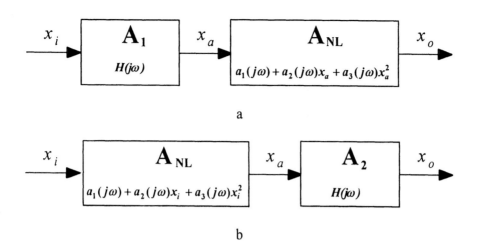

Fig. 7.3a. Amplifier models with: linear input stage and nonlinear output stage a), nonlinear input stage followed by a linear stage b).

Evaluation of harmonic distortion factors for the amplifier schematised in Fig. 7.3a is straightforward. We can use (7.7) and (7.8) after noting that the input signal of the nonlinear block is now $H(j\omega)X_i$. Hence, the distortion factors are

$$HD_2(\omega) \approx \frac{1}{2}\frac{|a_2(j\omega)|}{|a_1(j\omega)|}|H(j\omega)|X_i = \frac{1}{2}\frac{|a_2(j\omega)|}{|a_1(j\omega)|^2}|X_o(j\omega)| \qquad (7.37)$$

$$HD_3(\omega) \approx \frac{1}{4}\frac{|a_3(j\omega)|}{|a_1(j\omega)|}|H(j\omega)|^2 X_i^2 = \frac{1}{4}\frac{|a_3(j\omega)|}{|a_1(j\omega)|^3}|X_o(j\omega)|^2 \qquad (7.38)$$

Harmonic distortion referred to the amplitude of the output signal fundamental component are formally identical to the last equations in (7.7) and (7.8) except that now these expressions must be evaluated at the frequency of the input signal (fundamental). Note that this also holds for X_o, i.e., also the output signal must be calculated at the fundamental frequency. Consequently, when we have to evaluate the frequency behaviour of HD_2 and HD_3, it is easier to refer to their formulations in terms of the input signal X_i.

The above equations give the magnitude of HD_2 and HD_3, as this is the most common information required by designers. However, in their general form these equations can be used to obtain also information on phase distortion. In the following we will consider only the magnitude of distortion factors.

The second basic case considered is that of a nonlinear amplifier followed by a linear stage, as shown in Fig. 7.3b. Assuming, as usual, that the incremental input voltage is a pure sinusoidal tone, $x_i = X_i \cos(\omega t)$, the intermediate output is

$$x_a \approx |b_1(j\omega)|\cos(\omega t) + |b_2(j\omega)|\cos(2\omega t) + |b_3(j\omega)|\cos(3\omega t) \qquad (7.39)$$

where coefficients $b_i(j\omega)$ are again given by (7.3)-(7.6) in which $a_i(j\omega)$ have to be used instead of constant values a_i. Then, the output signal is

$$x_o \approx |b_1(j\omega)||H(j\omega)|\cos(\omega t) + |b_2(j\omega)||H(j2\omega)|\cos(2\omega t) + \qquad (7.40)$$

$$+ |b_3(j\omega)||H(j3\omega)|\cos(3\omega t)$$

In the above equations the phase contribution of $b_i(j\omega)$ to the x_a components and that of $H(j\omega)$ to x_o has been neglected. Finally, from (7.7) and (7.8) we get

$$HD_2(\omega) \approx \frac{1}{2}\frac{|a_2(j\omega)|}{|a_1(j\omega)|}\frac{|H(j2\omega)|}{|H(j\omega)|}X_i = \frac{1}{2}\frac{|a_2(j\omega)|}{|a_1(j\omega)|^2}\frac{|H(j2\omega)|}{|H(j\omega)|^2}|X_o(j\omega)| \quad (7.41)$$

$$HD_3(\omega) = \frac{1}{4}\frac{|a_3(j\omega)|}{|a_1(j\omega)|}\frac{|H(j3\omega)|}{|H(j\omega)|}X_i^2 = \frac{1}{4}\frac{|a_3(j\omega)|}{|a_1(j\omega)|^3}\frac{|H(j3\omega)|}{|H(j\omega)|^3}|X_o(j\omega)|^2 \quad (7.42)$$

Comparing the above expressions with (7.7) we see that the harmonic distortion factors are now multiplied by the ratio of the transfer function magnitudes evaluated at the frequency of the considered harmonics and at the fundamental frequency.

As a particular case, assume that coefficients a_i are constant, and that the transfer function $H(j\omega)$ has a single pole (the pole of the amplifier and also of the loop gain)

$$H(j\omega) = \frac{h}{1 + j\dfrac{\omega}{\omega_c}} \qquad (7.43)$$

Accordingly, (7.41) and (7.42) become

$$HD_2(\omega) = \frac{1}{2}\frac{a_2}{a_1}\left|\frac{1+j\omega/\omega_c}{1+j2\omega/\omega_c}\right|X_i = \frac{1}{2}\frac{a_2}{ha_1^2}\frac{\left|1+j\omega/\omega_c\right|^2}{\left|1+j2\omega/\omega_c\right|}\left|X_o(j\omega)\right| \quad (7.44)$$

$$HD_3(\omega) = \frac{1}{4}\frac{a_3}{a_1}\left|\frac{1+j\omega/\omega_c}{1+j3\omega/\omega_c}\right|X_i^2 = \frac{1}{4}\frac{a_3}{h^2a_1^3}\frac{\left|1+j\omega/\omega_c\right|^3}{\left|1+j3\omega/\omega_c\right|}\left|X_o(j\omega)\right|^2 \quad (7.45)$$

The above equations show reduction of the second and third harmonic distortion factors with respect to their low-frequency values. Indeed, at frequencies respectively equal to $\omega_c/2$ and $\omega_c/3$, the asymptotic diagrams of $HD_2(\omega)$ and $HD_3(\omega)$ start to linearly decrease. Then, the distortion factors become constant at the cut-off frequency. This behaviour is qualitatively shown in Fig. 7.4.

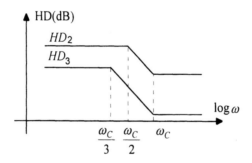

Fig. 7.4. Harmonic distortion factors for the scheme in Fig. 7.3b. Coefficients a_i are assumed constant and ω_c is the pole of $H(j\omega)$.

7.2.2 Closed-loop Amplifiers

Consider now the same feedback amplifier in Fig. 7.2, but where the transfer functions of blocks A_{NL} and f are now frequency dependent. Specifically, let block A_{NL} be characterised by the frequency-dependent nonlinear coefficients $a_1(j\omega)$, $a_2(j\omega)$, and $a_3(j\omega)$, and denote as $F(j\omega)$ the transfer function of linear feedback block f, as schematised in Fig. 7.5.

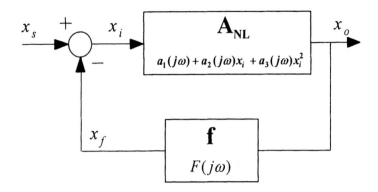

Fig. 7.5. Nonlinear feedback amplifier in the frequency domain.

To derive the distortion factors of the system in Fig. 7.5, we will now develop an intuitive method which requires simple algebraic manipulations. The approach leads to expressions of distortion factors that are a direct extension of those in (7.16) (7.17) found at low frequency.

As usual, we assume a sinusoidal input tone $x_s = X_s \cos(\omega t)$ and that it is possible to write the output signal as a power series of the source signal

$$x_o \approx \left|a_1'(j\omega)\right| X_s \cos\left(\omega t + \angle a_1'(j\omega)\right) + \frac{1}{2}\left|a_2'(j\omega)\right| X_s^2 \cos\left(2\omega t + \angle a_2'(j\omega)\right)$$

$$+ \frac{1}{4}\left|a_3'(j\omega)\right| X_s^3 \cos\left(3\omega t + \angle a_3'(j\omega)\right) \qquad (7.46)$$

The problem is to find the expression of the closed-loop nonlinear coefficients $a_i'(j\omega)$.

The first coefficient $a_1'(j\omega)$, which is responsible for the linear behaviour, can be simply found. It is equal to the forward-path transfer function divided by 1 plus the loop-gain transfer function, $T(j\omega)$

$$a_1'(j\omega) = \frac{a_1(j\omega)}{1 + a_1(j\omega)F(j\omega)} = \frac{a_1(j\omega)}{1 + T(j\omega)} \qquad (7.47)$$

Equation (7.47) implies computation of $T(j\omega)$ and $a_1(j\omega)$ at the frequency of the input tone (i.e., the fundamental frequency).

To evaluate the higher-order coefficients we have to follow a simple, but not trivial reasoning. Concentrate our attention to derive the second harmonic component at the output. It is produced by the nonlinear block when a signal at the fundamental frequency is presented to its input. Now observe that the second harmonics is proportional to x_i^2. If the circuit is perfectly linear (i.e., $a_2(j\omega) = a_3(j\omega) = 0$), x_i would be equal to $x_s/|1+T(j\omega)|$. Therefore, the nonlinear block produces a second harmonic component with amplitude equal to $\dfrac{1}{2}|a_2(j\omega)|\dfrac{1}{|1+T(j\omega)|^2}X_s^2$. This component can be viewed as a spurious signal injected at the output of the nonlinear block, as depicted in Fig. 7.6. It is subsequently processed by the feedback loop and appears at the output terminal decreased by the loop gain but evaluated at the *frequency of the harmonic considered*, i.e., 2ω.

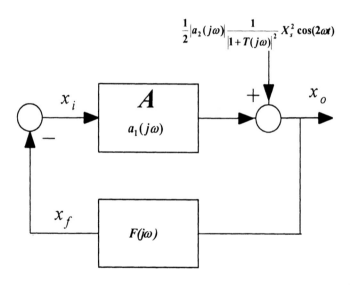

Fig. 7.6. Equivalent representation to evaluate second harmonic distortion in the frequency domain.

From the above discussion it follows that the nonlinear term $a'_2(j\omega)$ is equal to

$$a'_2(j\omega) = \frac{1}{[1+T(j\omega)]^2}\frac{1}{1+T(j2\omega)}a_2(j\omega) \qquad (7.48)$$

The nonlinear coefficient $a'_3(j\omega)$ can be evaluated by following a similar procedure. Neglecting the contribute due to $a_2(j\omega)$ we get

$$a'_3(j\omega) = \frac{1}{[1+T(j\omega)]^3} \frac{1}{1+T(j3\omega)} a_3(j\omega) \qquad (7.49a)$$

Taking into account also the effect of $a_2(j\omega)$ an expression similar to (7.15) can be deduced. At this purpose, we consider the schematisation depicted in Fig. 7.7 which leads to

$$a'_3(j\omega) \approx \frac{1}{[1+T(j\omega)]^3} \frac{1}{1+T(j3\omega)} a_3(j\omega)\left[1-\frac{2a_2^2(j\omega)}{a_1(j\omega)a_3(j\omega)}\right] \qquad (7.49b)$$

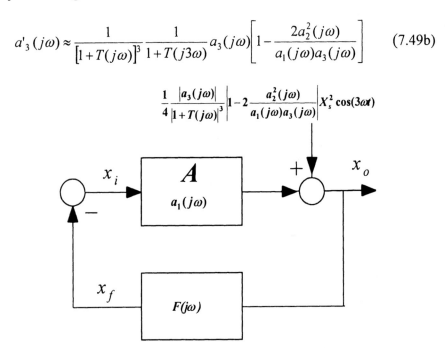

Fig. 7.7. Equivalent representation to evaluate complete third harmonic distortion in the frequency domain.

Substituting (7.47), (7.48) and (7.49b) into (7) and (8) we get

$$HD_{2f}(\omega) = \frac{1}{2}\frac{|a_2(j\omega)|}{|a_1(j\omega)|}\frac{1}{|1+T(j\omega)||1+T(j2\omega)|}X_s =$$

$$\qquad\qquad\qquad\qquad\qquad\qquad\qquad\qquad\qquad (7.50)$$

$$= \frac{1}{2}\frac{|a_2(j\omega)|}{|a_1(j\omega)|^2}\frac{1}{|1+T(j2\omega)|}|X_o(j\omega)|$$

$$HD_{3f}(\omega) = \frac{1}{4}\frac{|a_3(j\omega)|}{|a_1(j\omega)|}\frac{1}{|1+T(j\omega)|^2|1+T(j3\omega)|}\left|1-2\frac{a_2^2(j\omega)}{a_1(j\omega)a_3(j\omega)}\right|X_s^2$$

$$= \frac{1}{4}\frac{|a_3(j\omega)|}{|a_1(j\omega)|^3}\frac{1}{|1+T(j3\omega)|}\left|1-2\frac{a_2^2(j\omega)}{a_1(j\omega)a_3(j\omega)}\right||X_o(j\omega)|^2$$

$$(7.51)$$

Of course, the above equations adhere with (7.16), (7.17) and (7.19a) found in the case of frequency-independent loop gain, or that is the same, for low-frequency input signals. In the present case, distortion of a feedback network in terms of the output fundamental is reduced of a quantity still equal to the return ratio but evaluated at the *considered* harmonic.

It is useful to extend these results to a more general model in which we put the nonlinear block between two linear blocks in the forward path, as shown in Fig. 7.8a. This system includes as particular cases the closed-loop version of both occurrences, depicted in Fig. 7.3a and 7.3b, in which distortion appears after or before a linear stage.

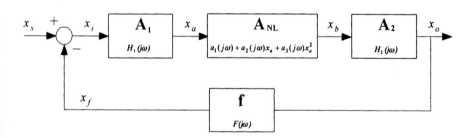

Fig. 7.8. A general model of closed-loop nonlinear amplifier for evaluation of harmonic distortion in the frequency domain.

To obtain distortion factors of the system in Fig. 7.8, we can follow the same procedure described above. Let us first evaluate the nonlinear coefficients that relate x_o to x_s. The first-order coefficient is

$$a'_1(j\omega) = \frac{H_1(j\omega)H_2(j\omega)}{1+T(j\omega)}a_1(j\omega)$$

$$(7.52)$$

where $T(j\omega) = a_1(j\omega)F(j\omega)H_1(j\omega)H_2(j\omega)$.

To obtain the second-order coefficient it is convenient to refer to Fig. 7.9, which illustrates the second-order component injected at the output of the nonlinear block

$$a'_2(j\omega) = \frac{H_1(j\omega)^2}{[1+T(j\omega)]^2} \frac{H_2(j2\omega)}{1+T(j2\omega)} a_2(j\omega) \tag{7.53}$$

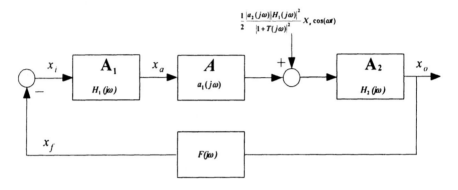

Fig. 7.9. Equivalent representation to evaluate the second harmonic distortion for system in Fig. 7.8.

A similar procedure can be applied to the third-order coefficient, yielding

$$a'_3(j\omega) = \left[\frac{H_1(j\omega)}{1+T(j\omega)}\right]^3 \frac{H_2(j3\omega)}{1+T(j3\omega)} a_3 \left[1 - \frac{2a_2^2(j\omega)}{a_3(j\omega)a_1(j\omega)}\right] \tag{7.54}$$

Then, the harmonic distortion factors are expressed by

$$HD_{2f}(\omega) = \frac{1}{2} \frac{|a_2(j\omega)|}{|a_1(j\omega)|} \frac{|H_1(j\omega)||H_2(j2\omega)|}{|1+T(j\omega)||1+T(j2\omega)||H_2(j\omega)|} X_s =$$

$$\tag{7.55}$$

$$= \frac{1}{2} \frac{|a_2(j\omega)|}{|a_1(j\omega)|^2} \frac{|H_2(j2\omega)|}{|1+T(j2\omega)||H_2(j\omega)|^2} |X_o(j\omega)|$$

$$HD_{3f}(\omega) = \frac{1}{4} \frac{|a_3(j\omega)|}{|a_1(j\omega)|} \frac{|H_1(j\omega)|^2 |H_2(j3\omega)|}{|1+T(j\omega)|^2 |1+T(j3\omega)||H_2(j\omega)|} \cdot \left|1 - 2\frac{a_2^2(j\omega)}{a_3(j\omega)a_1(j\omega)}\right| X_s^2$$

$$= \frac{1}{4} \frac{|a_3(j\omega)|}{|a_1(j\omega)|^3} \frac{|H_2(j3\omega)|}{|1+T(j3\omega)||H_2(j\omega)|^3} \cdot \left|1 - 2\frac{a_2^2(j\omega)}{a_3(j\omega)a_1(j\omega)}\right| |X_o(j\omega)|^2$$

$$(7.56)$$

7.3 HARMONIC DISTORTION AND COMPENSATION

In this paragraph we will study the effect of the different types of frequency compensation on harmonic distortion. To this end, we will first apply the above results to two-stage amplifiers and then a typical single-stage amplifier will be considered. Dominant-pole and Miller techniques for a two-stage amplifier are treated in sections 7.3.1 and 7.3.2, respectively. Under the assumption that the second stage is the principal responsible for nonlinear behaviour, we will demonstrate the better linearity performance of Miller-compensated amplifiers. Linearity performance of a single-stage architecture with dominant-pole compensation will be treated in section 7.3.3.

Linear and, unless specified, frequency-independent feedback is thorough considered for simplicity.

7.3.1 Two-stage Amplifier with Dominant-Pole Compensation

The analysis carried out in the previous paragraph can be now directly applied to two-stage amplifiers compensated with the dominant-pole technique.

We can use the same model in Fig. 7.8, and assume $H_2(j\omega) = 1$ and $H_1(j\omega)$ to be a single-pole transfer function given by (7.43) and here reported again

$$H_1(j\omega) = \frac{h}{1 + j\omega/\omega_c} \qquad (7.57)$$

This means that the nonlinearity is caused by the second stage. Assume also for simplicity the nonlinear coefficients a_i being independent of frequency.

From (7.50) and (7.51), being $T_o = hfa_1$ and $\omega_{GBW} = (1+T_o)\omega_c$, we get immediately

$$HD_{2f}(\omega) = \frac{1}{2}\frac{a_2}{a_1}\frac{h}{(1+T_o)^2}\frac{\left|1+j2\dfrac{\omega}{\omega_c}\right|}{\left|1+j\dfrac{2\omega}{\omega_{GBW}}\right|\left|1+j\dfrac{\omega}{\omega_{GBW}}\right|}X_s$$

$$(7.58)$$

$$=\frac{1}{2}\frac{a_2}{a_1^2}\frac{1}{(1+T_o)}\frac{\left|1+j2\dfrac{\omega}{\omega_c}\right|}{\left|1+j\dfrac{2\omega}{\omega_{GBW}}\right|}\left|X_o(j\omega)\right|$$

$$HD_{3f}(\omega) = \frac{1}{4}\frac{a_3}{a_1^3}\left(1-2\frac{a_2^2}{a_3 a_1}\right)\frac{h^2}{(1+T_o)^3}\frac{\left|1+j3\dfrac{\omega}{\omega_c}\right|}{\left|1+j\dfrac{\omega}{\omega_{GBW}}\right|\left|1+j\dfrac{3\omega}{\omega_{GBW}}\right|}X_s$$

$$(7.59)$$

$$=\frac{1}{4}\frac{a_3}{a_1^3}\left(1-2\frac{a_2^2}{a_3 a_1}\right)\frac{1}{(1+T_o)}\frac{\left|1+j3\dfrac{\omega}{\omega_c}\right|}{\left|1+j\dfrac{3\omega}{\omega_{GBW}}\right|}\left|X_o(j\omega)\right|^2$$

Second- and third-order harmonic distortion factors start to *linearly* increase (from their low-frequency values) at a frequency equal to $\omega_c/2$ and $\omega_c/3$, respectively. Moreover, they become constant at frequencies equal to $\omega_{GBW}/2$ and $\omega_{GBW}/3$, respectively. At ω_{GBW} they begin to decrease.

A final observation concerns the distortion caused by the first amplifier stage. Nonlinear contributions of the input stage are reduced by the loop gain at low frequencies, and by the compensation capacitor at high frequencies (compensation tends to shunt the output of the first stage). Therefore, assuming the output stage as a principal source of nonlinearity is very well justified both for low and high frequencies. We will show in the following that this assumption is inadequate for Miller-compensated amplifiers.

7.3.2 Two-stage Amplifier with Miller Compensation

Another important case study is the evaluation of distortion for a two-stage Miller-compensated amplifier. Let us first analyse the open-loop amplifier in Fig. 7.10a, in which the second stage is nonlinear. In the figure, R and v_1 are the output resistance and the output voltage of the first stage, whose transconductance is G_m. The second stage, instead, is modeled by a voltage-controlled voltage source, A_{NL}, to preserve simplicity. To this end, we can also model the first stage with its Thévenin equivalent. The open-loop output voltage is then expressed by

$$v_{out} = A_{NL}(v_1) = -\left(a_1 + a_2 v_1 + a_3 v_1^2\right)v_1 \qquad (7.60)$$

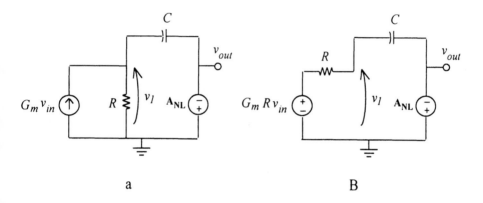

a   B

Fig. 7.10. Models of a Miller-compensated two-stage amplifier with a nonlinear output stage: Norton equivalent input stage a), Thévenin equivalent input stage.

The return ratio and the asymptotic gain of the amplifier in Fig. 7.10 are

$$T(j\omega) = \frac{j\omega RC}{1 + j\omega RC} a_1 \qquad (7.61)$$

$$G_\infty(j\omega) = -\frac{1}{j\omega RC} \qquad (7.62)$$

Given the Miller effect, we can consider the pole as being placed at the output of the first stage. Thus to analyse the circuit, we can use an equivalent block diagram similar to the one in Fig. 7.8 and depicted in Fig. 7.11 in which the nonlinear amplifier A_{NL} is characterised by the same nonlinear coefficients in (7.60).

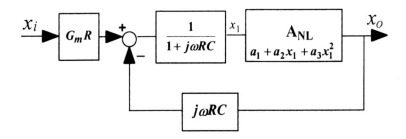

Fig. 7.11. Equivalent block diagram of the amplifier in Fig. 7.10b.

By comparing the general model in Fig. 7.8 with that in Fig. 7.11 we can utilise (7.52)-(7.54) to obtain the expression of the nonlinear closed-loop coefficients, where $H_2(j\omega) = 1$, $1/f = |G_\infty|$ and

$$H_1(j\omega) = \frac{G_\infty(j\omega)T(j\omega)}{a_1} = \frac{1}{1 + j\omega RC} \tag{7.63}$$

In addition, the closed-loop gain results

$$A_f(j\omega) = \frac{G_m R a_1}{1 + j\omega RC(1 + a_1)} \tag{7.64}$$

which, despite the different sign (inessential in evaluating distortion), equals the transfer function obtained by a direct inspection of the circuit in Fig. 7.10. Then, from relationship (7.52)-(7.54) we get the equivalent nonlinear coefficients $a'_1(j\omega)$, $a'_2(j\omega)$, and $a'_3(j\omega)$ which relate x_o to x_i in Fig. 7.11

$$a'_1(j\omega) = \frac{1}{1 + j\omega RC(1 + a_1)} a_1 \tag{7.65}$$

$$a'_2(j\omega) = \frac{1 + j2\omega RC}{[1 + j\omega RC(1 + a_1)]^2 [1 + j2\omega RC(1 + a_1)]} a_2 \tag{7.66}$$

$$a'_3(j\omega) = \frac{1 + j3\omega RC}{[1 + j\omega RC(1 + a_1)]^3 [1 + j3\omega RC(1 + a_1)]} a_3 \left(1 - 2\frac{a_2^2}{a_1 a_3}\right) \tag{7.67}$$

The closed-loop Miller-compensated amplifier can then be modeled as depicted in Fig. 7.12, where the amplifier studied above is closed in a loop with feedback block f. Note that to further simplify the scheme, Fig. 6.12b includes the new nonlinear block A'_{NL}, with its nonlinear coefficients $a'_1(j\omega)$, $a'_2(j\omega)$, and $a'_3(j\omega)$ defined above. Moreover, for conformity with the notation used in the previous section, we define the gain of the first block, h, as equal to $G_m R$.

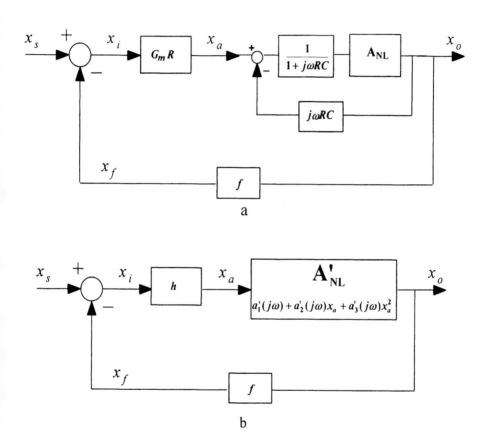

a

b

Fig. 7.12. Two-stage Miller-compensated closed-loop amplifier models.

Gain h in Fig. 7.12b is equal to $G_m R$ and coefficients a'_i are defined in

(7.65)-(7.67).

From relationships (7.50)-(7.51a), and given that $T_o = h f a_1$, we get the second and third harmonic distortion factors

$$HD_{2f}(\omega) = \frac{1}{2}\frac{a_2}{a_1}\frac{h}{(1+T_o)^2}\frac{|1+j2\omega RC|}{\left|1+j2\omega RC\dfrac{1+a_1}{1+T_o}\right|\left|1+jRC\dfrac{1+a_1}{1+T_o}\right|}X_s =$$

$$(7.68)$$

$$= \frac{1}{2}\frac{a_2}{a_1^2}\frac{1}{1+T_o}\frac{|1+j2\omega RC|}{\left|1+j2\omega RC\dfrac{1+a_1}{1+T_o}\right|}|X_o(j\omega)|$$

$$HD_{3f}(\omega) \approx \frac{1}{4}\frac{a_3}{a_1}\frac{h^2}{(1+T_o)^3}\left(1-2\frac{a_2^2}{a_1a_3}\right)\frac{|1+j3\omega RC|}{\left|1+j3\omega RC\dfrac{1+a_1}{1+T_o}\right|\left|1+j\omega RC\dfrac{1+a_1}{1+T_o}\right|^2}X_s^2 =$$

$$= \frac{1}{4}\frac{a_3}{a_1^3}\frac{1}{1+T_o}\left(1-2\frac{a_2^2}{a_1a_3}\right)\frac{|1+j3\omega RC|}{\left|1+j3\omega RC\dfrac{1+a_1}{1+T_o}\right|}|X_o(j\omega)|^2 \qquad (7.69)$$

where in (7.69) we have only considered the dominant terms.

To better compare the above results with those obtained in the case of dominant pole compensation we must express (7.68)-(7.69) in terms of ω_c, that is now equal to $1/[RC(1+a_1)]$ and ω_{GBW} equal to $(1+T_o)\omega_c$

$$HD_{2f}(\omega) = \frac{1}{2}\frac{a_2}{a_1}\frac{h}{(1+T_o)^2}\frac{\left|1+j2\dfrac{\omega}{(1+a_1)\omega_c}\right|}{\left|1+j2\dfrac{\omega}{\omega_{GBW}}\right|\left|1+j\dfrac{\omega}{\omega_{GBW}}\right|}X_s =$$

$$(7.70)$$

$$= \frac{1}{2}\frac{a_2}{a_1^2}\frac{1}{1+T_o}\frac{\left|1+j2\dfrac{\omega}{(1+a_1)\omega_c}\right|}{\left|1+j2\dfrac{\omega}{\omega_{GBW}}\right|}|X_o(j\omega)|$$

$$HD_{3f}(\omega) \approx \frac{1}{4}\frac{a_3}{a_1}\frac{h^2\left(1-2\frac{a_2^2}{a_1 a_3}\right)}{(1+T_o)^3}\frac{\left|1+j3\frac{\omega}{(1+a_1)\omega_c}\right|}{\left|1+j3\frac{\omega}{\omega_{GBW}}\right|\left|1+j\frac{\omega}{\omega_{GBW}}\right|^2}X_s^2 =$$

(7.71)

$$= \frac{1}{4}\frac{a_3}{a_1^3}\frac{1-2\frac{a_2^2}{a_1 a_3}}{1+T_o}\frac{\left|1+j3\frac{\omega}{(1+a_1)\omega_c}\right|}{\left|1+j3\frac{\omega}{\omega_{GBW}}\right|}\left|X_o(j\omega)\right|^2$$

Starting from their low-frequency values, second- and third-order harmonic distortion factors linearly increase at a frequency equal to $(1+a_1)\omega_c/2$ and $(1+a_1)\omega_c/3$, respectively. Compared to dominant-pole compensation, we see that the frequency band where distortion factors remain equal to their low-frequency values is greater in the Miller-compensated amplifier by a factor equal to $1+a_1$.

Equations (7.70) and (7.71) also predict that HD_{2f} and HD_{3f} become constant at frequencies equal to $\omega_{GBW}/2$ and $\omega_{GBW}/3$, respectively. At ω_{GBW} they begin to decrease. This behaviour was already found appropriate in two-stage amplifiers compensated with a dominant pole. In contrast, when using Miller compensation it is unrealistic. Indeed, the local feedback operated by the Miller capacitor causes coefficients $a_i'(j\omega)$ to decrease with frequency. At high frequencies, distortion of the first stage becomes dominant and a nonlinear model of the first stage should then be included to accurately predict harmonic distortion.

The use of nonlinear models for both the first and second stage considerably complicates distortion evaluation. However, since the two distortion mechanisms are dominant over different frequency ranges (distortion due to the input stage is effective at high frequencies, whilst distortion due to the output stage is dominant at low frequencies) we can separately study the two cases with our distortion models[4]. We shall not use this approach now, because it can be shown that fairly good approximation for distortion factors valid up to the gain-bandwidth product is found simply

---

[4] An example of how to treat distortion coming from two cascaded stages is described in the next section, 7.3.2.

by eliminating the poles in (7.70) and (7.71) respectively at ω_{GBW} , $\omega_{GBW}/2$ and at ω_{GBW} , $\omega_{GBW}/3$.

As a result, HD_{2f} and HD_{3f} for a two-stage amplifier compensated with Miller technique are expressed by

$$HD_{2f}(\omega) \approx \frac{1}{2}\frac{a_2}{a_1}\frac{h}{(1+T_o)^2}\left|1+j2\frac{\omega}{(1+a_1)\omega_c}\right|X_s =$$

$$(7.72)$$

$$=\frac{1}{2}\frac{a_2}{a_1^2}\frac{1}{1+T_o}\left|1+j2\frac{\omega}{(1+a_1)\omega_c}\right|\left|1+j\frac{\omega}{\omega_{GBW}}\right|X_o(j\omega)$$

$$HD_{3f}(\omega) \approx \frac{1}{4}\frac{a_3}{a_1}\frac{h^2\left(1-2\dfrac{a_2^2}{a_1 a_3}\right)}{(1+T_o)^3}\left|1+j3\frac{\omega}{(1+a_1)\omega_c}\right|X_s^2 =$$

$$(7.73)$$

$$=\frac{1}{4}\frac{a_3}{a_1^3}\frac{1-2\dfrac{a_2^2}{a_1 a_3}}{1+T_o}\left|1+j3\frac{\omega}{(1+a_1)\omega_c}\right|\left|1+j\frac{\omega}{\omega_{GBW}}\right|^2 |X_o(j\omega)|^2$$

To qualitatively illustrate the improvement in linearity of Miller compensation over dominant-pole compensation, Figure 7.13 shows the HD_{2f} achieved in both cases. A similar plot can be drawn for HD_{3f}

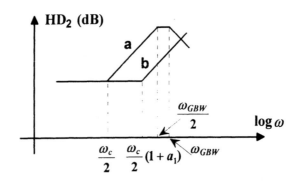

Fig. 7.12. Typical behaviour of second-order distortion factors for two-stage amplifiers with dominant-pole compensation (curve a) and Miller compensation (curve b).

7.3.3 Single-stage Amplifiers

The last case we shall study is that of the single-stage amplifiers. These architectures are frequently employed in IC applications (for instance in switched-capacitor circuits) for their high-frequency performance. Indeed, a single-stage amplifier exhibits only an (output) high-resistance node. Moreover, this output node often exploits cascoding, allowing a voltage gain similar to that of two-stage amplifiers to be achieved. Of course, these amplifiers are used in closed-loop configurations and, due to the internal structure, output dominant-pole compensation is invariably utilised.

The small-signal model of a (open-loop) single-stage amplifier is illustrated in Fig. 7.13, in which C is the output compensation capacitor.

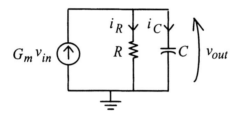

Fig. 7.13. Small-signal model of a single-stage amplifier.

In general, there are two sources of harmonic distortion in such amplifiers. The first is due to the nonlinear V-I conversion accomplished by the input transconductance stage. The second is due to the nonlinear I-V characteristic exhibited by the output devices.

Let us first analyse the effect on linearity of the nonlinear output resistance. Observe that this case does not fall into the category of any of those already studied because both pole and distortion are generated at the same circuit node (i.e., the output) by the same nonlinear element. Hence a specific analysis must be performed.

For easy calculation express the input signal as $v_{in} = V_M e^{j\omega t}$. Moreover, it is better to characterise the nonlinear resistance in terms of (nonlinear) conductance

$$i_R = \frac{1}{R}\left(1 + g_{2N}v_{out} + g_{3N}v_{out}^2\right)v_{out} \tag{7.72}$$

where g_{2N} and g_{3N} are nonlinear coefficients normalised to the linear part of the output conductance $1/R$. These cause harmonic distortion components to appear in the output voltage, according to

$$v_{out} = b_1(j\omega) \cdot V_M e^{j\omega t} + b_2(j\omega) \cdot V_M^2 e^{j2\omega t} + b_3(j\omega) \cdot V_M^3 e^{j3\omega t} \qquad (7.73)$$

in which only the first three terms are taken. Then, the current through the capacitor is

$$i_C = C\frac{dv_{out}}{dt} = \qquad\qquad\qquad\qquad\qquad (7.74)$$

$$= C\left[j\omega b_1(j\omega) \cdot V_M e^{j\omega t} + j2\omega b_2(j\omega) \cdot V_M^2 e^{j2\omega t} + j3\omega b_3(j\omega) \cdot V_M^3 e^{j3\omega t} \right]$$

From the KCL at the output node

$$G_m v_{in} = i_R + i_C \qquad\qquad\qquad\qquad\qquad (7.75)$$

using (7.74) and the current through the nonlinear resistor found by substituting (7.73) in (7.72), and equating all the harmonic components with the same frequency, we can derive the expression of coefficients $b_1(j\omega)$, $b_2(j\omega)$, and $b_3(j\omega)$. Thus, considering only the dominant terms we get

$$b_1(j\omega) = \frac{G_m R}{1 + j\omega RC} \qquad\qquad\qquad\qquad (7.76)$$

$$b_2(j\omega) = -\frac{g_{2N}(G_m R)^2}{(1 + j\omega RC)^2 (1 + j2\omega RC)} \qquad\qquad (7.77)$$

$$b_3(j\omega) = -\frac{(G_m R)^3}{(1 + j\omega RC)^3 (1 + j3\omega RC)}\left(g_{3N} - \frac{2g_{2N}^2}{1 + j2\omega RC} \right) \qquad (7.78)$$

Normalising the second and third coefficient to $b_1(j\omega)$, and given that $\omega_c = 1/RC$, we get

$$b_{2N}(j\omega) = \frac{b_2(j\omega)}{b_1(j\omega)} = -\frac{g_{2N} G_m R}{\left(1 + j\dfrac{\omega}{\omega_c}\right)\left(1 + j\dfrac{2\omega}{\omega_c}\right)} \qquad (7.79)$$

$$b_{3N}(j\omega) = \frac{b_3(j\omega)}{b_1(j\omega)} = -\frac{(G_m R)^2}{\left(1 + j\dfrac{\omega}{\omega_c}\right)^2 \left(1 + j\dfrac{3\omega}{\omega_c}\right)}\left(g_{3N} - \frac{2g_{2N}^2}{1 + j\dfrac{2\omega}{\omega_c}}\right) \quad (7.80)$$

Hence, the feedback circuit can then be schematised by the block diagram in Fig. 7.14, where the blocks inside the shadowed area represent the linear and nonlinear contributes of the RC output node, with the nonlinear coefficients given by

$$b_{2N}'(j\omega) = -\frac{g_{2N}}{\left(1 + j\dfrac{2\omega}{\omega_c}\right)} \quad (7.81)$$

$$b_{3N}'(j\omega) = -\frac{1}{\left(1 + j\dfrac{3\omega}{\omega_c}\right)}\left(g_{3N} - \frac{2g_{2N}^2}{1 + j\dfrac{2\omega}{\omega_c}}\right) \quad (7.82)$$

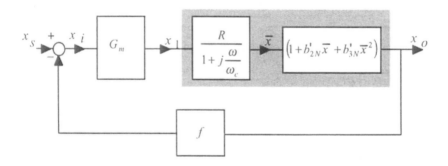

Fig. 7.14. Model of a single-stage amplifier with a nonlinear output resistance.

To evaluate the closed-loop harmonic distortion factors we can employ the results found at the end of section 7.2.2. After applying (7.55) and (7.56) we get the following equations in which $T_o = fG_m R$ and $\omega_{GBW} = (1 + T_o)\omega_c$ (expressions in terms of X_s only are reported for compactness).

$$HD_{2f}^{R}(\omega) = \frac{1}{2} \frac{G_m b_{2N}}{|1+T(j\omega)||1+T(j2\omega)|} X_s =$$

$$(7.83)$$

$$= \frac{1}{2} \frac{G_m R}{(1+T_o)^2} \frac{g_{2N}}{\left|1+j\dfrac{\omega}{\omega_{GBW}}\right|\left|1+j\dfrac{2\omega}{\omega_{GBW}}\right|} X_s$$

$$HD_{3f}^{R}(\omega) = \frac{1}{4} \frac{G_m^2 b_{3N}}{|1+T(j\omega)|^2 |1+T(j2\omega)|} X_s^2 =$$

$$(7.84)$$

$$= \frac{1}{4} \frac{(G_m R)^2}{(1+T_o)^3} \frac{g_{3N} - \dfrac{2g_{2N}^2}{1+j\dfrac{2\omega}{\omega_c}}}{\left|1+j\dfrac{\omega}{\omega_{GBW}}\right|^2 \left|1+j\dfrac{3\omega}{\omega_{GBW}}\right|} X_s^2$$

Distortion due to the nonlinear output conductance is effective at low frequencies. Indeed, so long as the loop gain is high, signal x_i (the *error signal*) is small, and distortion is mainly due to nonlinearities arising in the output resistance R which is operated under large-signal conditions. For increasing frequencies the compensation capacitor shunts the output impedance to ground thereby reducing the weight of nonlinearities due to the output resistance. Moreover, signal x_i increases (due to the reduction in the loop gain) and the nonlinear effects of the input transconductance become more pronounced. Thus at high frequencies the amplifier is more adequately modeled by the block diagram in Fig. 7.15, which includes normalised nonlinear coefficients of the input transconductance a_{2N} and a_{3N}, and assumes the output resistance to be linear.

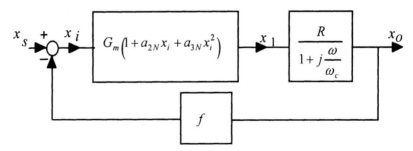

Fig. 7.15. Model of a single-stage amplifier with a nonlinear input transconductance.

This scheme is equivalent to the one analysed in Fig. 7.8 by properly updating the block transfer functions. Hence, utilising (7.55) and (7.56) we get

$$HD_{2f}^{Gm}(\omega) = \frac{1}{2} \frac{a_{2N}}{|1+T(j\omega)||1+T(j2\omega)| \left|\dfrac{R}{1+j\dfrac{\omega}{\omega_c}}\right|} \left|\frac{R}{1+j\dfrac{2\omega}{\omega_c}}\right| X_s =$$

(7.85)

$$= \frac{1}{2} \frac{a_{2N}}{(1+T_o)^2} \frac{\left|1+j\dfrac{\omega}{\omega_c}\right|^2}{\left|1+j\dfrac{\omega}{\omega_{GBW}}\right|\left|1+j\dfrac{2\omega}{\omega_{GBW}}\right|} X_s$$

$$HD_{3f}^{Gm}(\omega) = \frac{1}{4} \frac{a_{3N}\left(1-2\dfrac{a_{2N}^2}{a_{3N}}\right)}{|1+T(j\omega)|^2|1+T(j2\omega)| \left|\dfrac{R}{1+j\dfrac{\omega}{\omega_c}}\right|} \left|\frac{R}{1+j\dfrac{3\omega}{\omega_c}}\right| X_s^2 =$$

(7.86)

$$= \frac{1}{4} \frac{a_{3N}-2a_{2N}^2}{(1+T_o)^3} \frac{\left|1+j\dfrac{\omega}{\omega_c}\right|^3}{\left|1+j\dfrac{\omega}{\omega_{GBW}}\right|^2\left|1+j\dfrac{3\omega}{\omega_{GBW}}\right|} X_s^2$$

Both the above distortion factors increase for frequencies higher than the amplifier pole. As a consequence, their effects can be significant at high frequencies.

To qualitatively compare the effects on output distortion due to the output resistance and the input transconductance, let us consider the plots in Fig. 7.16. They illustrate the typical behaviour of second harmonic distortion factors due to the nonlinear output resistance, HD_{2f}^R, and due to the input transconductance, HD_{2f}^{Gm}. The frequency determining which contribution is dominant is located between ω_c and $\omega_{GBW}/2$ and is close to ω_c if $HD_{2f}^R(0)$ approaches $HD_{2f}^{Gm}(0)$.

Similar plots can also be deduced for the third harmonic distortion factors.

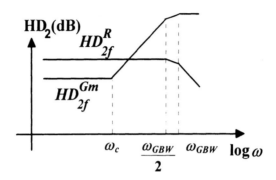

Fig. 7.16. Typical plots of second harmonic distortion factors due to the nonlinear output resistance, HD_{2f}^R, and the input transconductance, HD_{2f}^{Gm}.

As a final analysis step, we consider the two distortion mechanisms together in the same block scheme as depicted in Fig. 7.17

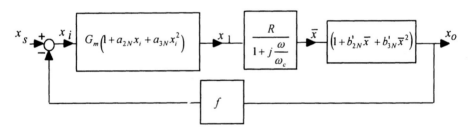

Fig. 7.17. Model of a single-stage amplifier with both nonlinear input transconductance and output resistance.

The exact resolution of this system is difficult, but can fortunately be avoided by considering that the two distortion mechanisms are dominant over different frequency ranges, as previously stated. Consequently, expressions of complete distortion factors HD_{2f} and HD_{3f} which provide asymptotic approximation can be found by combining (7.83) with (7.85) and (7.84) with (7.86)

$$HD_{2f}(\omega) = \frac{1}{2}\frac{G_m R}{(1+T_o)^2} \frac{\left| g_{2N} + \dfrac{a_{2N}}{G_m R}\left(1+j\dfrac{\omega}{\omega_c}\right)^2 \right|}{\left| 1+j\dfrac{\omega}{\omega_{GBW}} \right| \left| 1+j\dfrac{2\omega}{\omega_{GBW}} \right|} X_s$$

(7.87)

$$HD_{3f}(\omega) = \frac{1}{4}\frac{(G_m R)^2}{(1+T_o)^3} \frac{\left| g_{3N} - \dfrac{2g_{2N}^2}{1+j\dfrac{2\omega}{\omega_c}} + \dfrac{a_{3N}-2a_{2N}^2}{(G_m R)^2}\left(1+j\dfrac{\omega}{\omega_c}\right)^3 \right|}{\left| 1+j\dfrac{\omega}{\omega_{GBW}} \right|^2 \left| 1+j\dfrac{3\omega}{\omega_{GBW}} \right|} X_s$$

(7.88)

The above relationships have simply been obtained by algebraically adding, before taking their modules, HD_{2f}^R with HD_{2f}^{Gm} and HD_{3f}^R with HD_{3f}^{Gm}.

7.4 AN ALTERNATIVE FREQUENCY ANALYSIS

In this paragraph we describe a simple analytical procedure to calculate the closed-loop harmonic distortion factors in the frequency domain, already found in section 7.2.2 through an euristic demonstration, and used in this chapter. Refer again to Fig. 7.5 and express the source signal as

$$x_s = X_s e^{j\omega t}$$

(7.89)

Due to the nonlinear block in the direct path the output signal will include harmonic components. Assume it is given by

$$x_o = a_1'(j\omega)X_s e^{j\omega t} + a_2'(j\omega)X_s^2 e^{j2\omega t} + a_3'(j\omega)X_s^3 e^{j3\omega t}$$

(7.90)

where coefficients $a_1'(j\omega)$, $a_2'(j\omega)$, and $a_3'(j\omega)$ have to be determined.

The error signal, x_i, is the difference of the source signal and the output signal times the value of the feedback factor evaluated at the appropriate frequency

$$x_i = X_s e^{j\omega t} - a_1'(j\omega)X_s e^{j\omega t} F(j\omega) - a_2'(j\omega)X_s^2 e^{j2\omega t} F(j2\omega) -$$

$$- a_3'(j\omega)X_s^3 e^{j3\omega t} F(j3\omega)$$

$$(7.91)$$

then it is processed by the nonlinear block whose output is

$$x_o = X_s \left[1 - F(j\omega)a_1'(j\omega)\right] e^{j\omega t} a_1(j\omega) -$$

$$- F(j2\omega)a_2'(j\omega)X_s^2 e^{j2\omega t} a_1(j2\omega) - F(j3\omega)a_3'(j\omega)X_s^3 e^{j3\omega t} a_1(j3\omega) +$$

$$+ X_s^2 \left[1 - F(j\omega)a_1'(j\omega)\right]^2 e^{j2\omega t} a_2(j\omega) +$$

$$+ X_s^3 \left[1 - F(j\omega)a_1'(j\omega)\right]^3 e^{j3\omega t} a_3(j\omega)$$

$$(7.92)$$

After substituting (7.90) in (7.92) and equating the terms with the same frequency component in the exponential factor, we get

$$a_1'(j\omega) = \left[1 - F(j\omega)a_1'(j\omega)\right]a_1(j\omega)$$

$$(7.93)$$

$$a_2'(j\omega) = -F(j2\omega)a_2'(j\omega)a_1(j2\omega) + \left[1 - F(j\omega)a_1'(j\omega)\right]^2 a_2(j\omega)$$

$$(7.94)$$

$$a_3'(j\omega) = -F(j\omega)a_3'(j\omega)a_1(j3\omega) + \left[1 - F(j\omega)a_1'(j\omega)\right]^3 a_3(j\omega)$$

$$(7.95)$$

Solving the above system for $a_1'(j\omega)$, $a_2'(j\omega)$, and $a_3'(j\omega)$ yields the same results as in (7.47), (7.48) and (7.49a).

Chapter 8

NOISE

Electronic noise is caused by small spontaneous fluctuations of currents and voltages associated with circuit components. For this reason noise cannot be predicted exactly, nor completely eliminated, but only can be minimised. Under this definition we explicitly exclude all the disturbance and interference (e.g. electrostatic and electromagnetic couplings) coming from sources external to the system being studied, most of which are deterministic and can be completely eliminated by adequate shielding, filtering, or by changing the system physical location.

Noise in electronic circuits is originated by resistors (which generate thermal, or white, noise) and by active devices. For instance, bipolar transistors contain different sources of noise: thermal noise, 1/f noise, and shot noise [M88].

In this chapter, after recalling some basic definitions, we shall concentrate our study on the effect of noise in feedback amplifiers. We will show that the noise properties of closed-loop amplifiers are not influenced (in any sense) by feedback. However, an added feedback network, when made up of resistive elements, will add noise.

8.1 BASIC CONCEPTS

We shall now recall some definitions associated with noise. For a broader coverage of the subject of noise in electronic system design, the interested reader is referred to [MC93], [GM93], [L94], [LS94], and [F88].

Noise is random and its average value over a certain period of time T is zero. Consequently, it is characterised and measured in the *mean-square* or *root-mean-square* (rms). If we denote with $x(t)$ a generic time-dependent noise variable that can either be a voltage or a current, its mean-square value is symbolised by $\overline{x^2}$ and its rms value by $\bar{x} = \sqrt{\overline{x^2}}$. The rms definition is based on the equivalent heating effect

$$\bar{x} = \sqrt{\frac{1}{T}\int_0^T x^2(t)dt} \qquad (8.1)$$

For electronic circuits, rms noise voltages and currents are usually expressed in the nV_{rms} and pA_{rms} ranges, respectively.

The frequency spectrum of noise extends from nearly zero to frequencies up to 10^{14} Hz. However, it is measured by instruments with limited bandwidth. Therefore, it is often convenient to express noise and particularly its mean square value in a 1-Hz unit of bandwidth

$$S_X = \frac{\overline{x^2}}{\Delta f} \qquad (8.2)$$

Since power is proportional to the square of voltage (current), S_X is called the *power spectral density* (PSD) of x and measured in V^2/Hz (A^2/Hz). Note that the square root of S_X (symbolised by $\bar{x}/\sqrt{\Delta f}$) is also a quantity of interest and its unit is $V/\sqrt{Hz}$ ($A/\sqrt{Hz}$).

Spectral density is a narrowband noise. In order to obtain the *total* wideband noise, (8.2) can be used only if S_X is constant with frequency. Otherwise the general relation between $\overline{x^2}$ and $S_X(f)$ is

$$\overline{x^2} = \int_0^\infty S_X(f)df \qquad (8.3)$$

In evaluating the output noise due to a single noise source, the usual rules used for networks in a sinusoidal steady state apply.

In contrast, when we have different uncorrelated[1] noise sources, the output noise is calculated as the root of the sum of the mean square values of each component. To better illustrate this concept, let us consider a generic network containing L noise voltage sources and M noise current sources. Here the total PSD of the output noise, S_{out}, is

$$S_{out}(\omega) = \sum_{l}^{L} |H_l'(j\omega)|^2 S_{Vl}(f) + \sum_{m}^{M} |H_m''(j\omega)|^2 S_{Im}(f) \tag{8.4}$$

where $|H_l'(j\omega)|$ and $|H_m''(j\omega)|$ are the magnitudes of the voltage and current transfer functions respectively from the noise voltage and current sources to the specified output variable. Note that since these transfer functions depend on the angular frequency, ω, the total PSD of the output noise in (8.4) is also function of ω. Using (8.3) we can express the total output rms noise as

$$\bar{x} = \sqrt{\frac{1}{2\pi} \int_{0}^{\infty} S_{out}(\omega) d\omega} \tag{8.5}$$

where division by 2π is made necessary by the change in the integration variable from f to ω.

8.2 EQUIVALENT INPUT NOISE GENERATORS

Consider the noisy two-port linear network in Fig. 8.1a. In order to compare the noise generated by the network to the incoming signal (and to its associated noise) we define equivalent input-referred noise generators. These generators, when applied to the *same* network, but considered noiseless (i.e., without internal noise sources) will produce the *same* output noise. Specifically, we need one voltage generator in series with the input, a current generator in parallel with the input, and a correlation coefficient which can have any value between -1 and +1. The latter takes into account the presence of common phenomena that contributes to both the two generators. As stated in note 1 of this chapter, this correlation can be usually

---

[1]Two noise quantities are said to be uncorrelated if they are produced independently and there is no relationship between their instantaneous values. Under a wide variety of actual situations (including active devices and operational amplifiers) the correlation between the different noise sources is zero or can be neglected.

neglected, otherwise it is simpler to return to the original network with internal noise sources. If correlation is neglected, the noise model simplifies to only $\overline{v_n}$ and $\overline{i_n}$ as illustrated in Fig. 8.1b. Note the use of generator symbols with unspecified polarity and characterised by rms values.

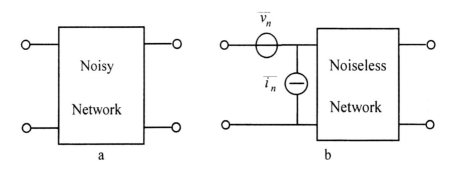

Fig. 8.1. Noisy two-port network a), equivalent noiseless network with input-referred noise generators b).

Now we explain how to evaluate these two equivalent sources. To calculate the equivalent input noise voltage, we equate the output noise of the original network to that of the noiseless network with input-referred noise generators, both under the condition of input shorted (thus the equivalent input noise current is shorted to ground and can be left out). Conversely, the equivalent input noise current can be found by equating the output noise of the two networks with the input left open (now it is the equivalent input noise voltage that can be left out). Note that for an N-input-port network we need, for each port, two noise generators, and noise is adequately modeled by $2N$ generators in total.

A particular example of a two-port network is an open-loop amplifier. Figure 82 shows a voltage amplifier, with gain A_v and input impedance z_{in}, and its associated equivalent input noise generators. An input signal source is connected at the input-port with a noisy series resistance R_S. This contribution is modeled with the voltage generator[2] $\overline{v_{Rs}}$.

We are now interested in the evaluation of the PSD of the output noise voltage, S_{Vout}, due to the combined action of the amplifier and the input source.

By defining the system gain as the gain from the input signal source to the output (different from the open-loop gain A_v)

---

[2] Noise models of circuit components are briefly described in paragraph 8.3.

$$\frac{v_{out}}{v_s} = \frac{z_{in}}{R_S + z_{in}} A_v \tag{8.6}$$

after simple calculations we get

$$S_{Vout} = \left|\frac{z_{in}}{R_s + z_{in}}\right|^2 |A_v|^2 \left(S_{VRs} + S_V\right) + \left(z_{in}\|R_s\right)^2 |A_v|^2 S_I \tag{8.7}$$

where S_V, S_I, and S_{Rs} are the PSD of $\overline{v_n}$, $\overline{i_n}$ and $\overline{v_{Rs}}$, respectively.

The above equation shows that both input referred generators are required to model the noise of the amplifier for any value of R_S. In fact, if we used only $\overline{v_n}$ ($\overline{i_n}$) no output noise due to the amplifier would be generated so long as R_S is zero (infinite). Therefore, the value of R_S determines which of the two noise generators is dominant.

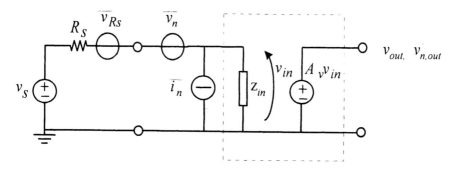

Fig. 8.2. Voltage amplifier with input-referred noise generators.

The PSD of the output noise divided by the square of the magnitude of the system gain gives the expression of the equivalent input noise voltage PSD, S_{Vin}

$$S_{Vin} = S_{VRs} + S_V + R_S^2 S_I \tag{8.8}$$

Thus, if the value of the source resistance is specified, the three noise sources may be modeled by only one noise generator of PSD S_{Vin}, as shown in Fig. 8.3

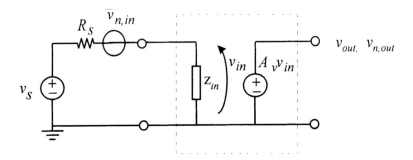

Fig. 8.3. Voltage amplifier with only one equivalent input noise voltage generator.

8.3 NOISE MODELS OF CIRCUIT COMPONENTS

In the previous paragraph we modeled the noise of an input signal source as a noise voltage associated to its series resistance. Actually, for noise calculations it is advantageous to have noise models for each active and passive circuit component. These models are derived in detail in standard textbooks like [GM93], here we summarise some results useful for our ensuing discussion.

Resistor
A resistance R generates thermal noise voltage whose PSD is given by

$$S_{VR} = 4kTR \qquad\qquad (8.9a)$$

in which k is the Boltzmann's constant and T is the absolute temperature ($kT = 41.4\cdot10^{-22}$ V^2/ΩHz). In Fig. 8.4a this noise is modeled by a voltage generator in series with the resistor. In some circuit configurations it is more convenient to represent this noise by a parallel current generator (see Fig. 8.4b) whose PSD is

$$S_{IR} = \frac{4kT}{R} \qquad\qquad (8.9b)$$

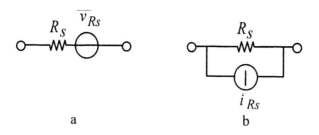

Fig. 8.4. Resistor noise models: voltage generator a), current generator b).

Transistor

Transistors contain several independent sources of noise. These sources can be referred to the input port (terminals Y-X of the generalised transistor) in order to be modeled by two noise generators according to the procedures described in the previous paragraph. Thus, the generalised transistor model introduced in Chapter 2 is now modified to include the equivalent input noise generators, $\overline{v_n}$ and $\overline{i_n}$, as depicted in Fig. 8.5

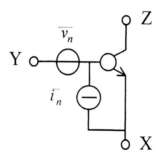

Fig. 8.5. Generic transistor with input-referred noise voltage and current generators.

It can be shown that noise magnitude depends on the transistor operating point, and that both the noise sources are important for BJTs, while the voltage source is dominant in FETs at low frequencies. However, the FET noise voltage is higher than that of the BJTs.

Operational Amplifier

Operational amplifiers can be arranged in inverting, noninverting, and differential closed-loop configurations. Thus all of these configurations must be adequately represented by an opamp noise model. Opamps exhibit two input ports and consequently they must be modeled by four noise generators, $\overline{v_{n1}}$, $\overline{v_{n2}}$, $\overline{i_{n1}}$, $\overline{i_{n2}}$, as illustrated in Fig. 8.6. Since virtually any opamp input stage is implemented by a differential amplifier which is characterised by a highly symmetrical topology, the rms values of the voltage (current) generators are equal.

If operated under negative feedback, the well-known virtual short circuit exists between the two input terminals. It allows the two noise voltage sources to be summed, indeed, they appear as connected in series. Therefore, the ultimate opamp noise model is that depicted in Fig. 8.6b in which the PSD is given by the sum of each component, $S_V = S_{V1} + S_{V2}$.

Finally, observe that in all those configurations in which the noninverting input terminal is grounded (e.g., the inverting opamp configuration), the noise current $\overline{i_{n2}}$ has no effects and can be left out of the circuit.

More details on operational amplifiers noise can be found in [TO89], [XDA00], [L95], and [A95].

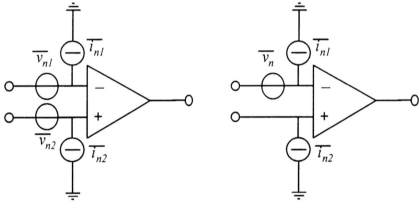

Fig. 8.6. Operational amplifier noise model: a) with two noise voltages, b) with one equivalent noise voltage.

8.4 EFFECT OF FEEDBACK

To see how noise is affected by feedback, consider the block diagram shown in Fig. 8.7, representing a two-stage amplifier. Gains H1 and H2 are the gain of the amplifier stages and f is the feedback factor. The input signal is v_s and the $\overline{v_{ni}}$ are noise voltages injected at various critical nodes, with PSD S_{Vi}.

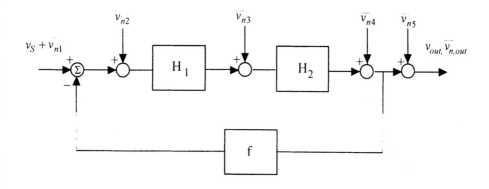

Fig. 8.7. Two-stage amplifier with noise voltage sources.

Let us first evaluate the total output noise of the amplifier without feedback (i.e., with $f = 0$). This is

$$S_{Vout}\big|_{ol} = S_{V5} + S_{V4} + S_{V3}H_2^2 + (S_{V1} + S_{V2})H_1^2 H_2^2 \tag{8.10}$$

Then evaluate the total output noise in closed-loop conditions. We get

$$S_{Vout}\big|_{cl} = S_{V5} + \frac{S_{V4}}{(1 + fH_1H_2)^2} + S_{V3}\frac{H_2^2}{(1 + fH_1H_2)^2} +$$

$$\tag{8.11}$$

$$+ (S_{V1} + S_{V2})\frac{H_1^2 H_2^2}{(1 + fH_1H_2)^2}$$

The output PSD noise voltage is lower in the latter case. Indeed, every component (except S_{V5}) is divided by the square of the loop gain. This is not a surprising result since the closed-loop gain is smaller than the open-loop gain by just the same quantity. Since the two systems have different gains, the output noise cannot be used for meaningful comparison. Instead, we can use the equivalent input noise for this purpose. It is found for the open- and closed-loop configuration by dividing the output PSD noise by the square of the system gain. We get the following results for the two cases

$$S_{Vin}\big|_{ol} = S_{V1} + S_{V2} + \frac{S_{V3}}{H_1^2} + \frac{S_{V4} + S_{V5}}{H_1^2 H_2^2} \tag{8.12a}$$

$$S_{Vin}\big|_{cl} = S_{V1} + S_{V2} + \frac{S_{V3}}{H_1^2} + \frac{S_{V4}}{H_1^2 H_2^2} + \frac{S_{V5}}{H_1^2 H_2^2}\left(1 + fH_1 H_2\right)^2 \qquad (8.12b)$$

Observe that each noise source, when referred to the input, is reduced by the (squared) gain between the input and the point where noise has been injected. This means that the principal noise contribution in cascaded amplifiers is due to the first stage, provided that the gain of this stage is sufficiently large. This fact also justifies the use of low-noise preamplifiers to improve the performance of noisy amplifiers (for example, in power audio amplifiers affected by power-supply hum). In addition, comparison of (8.12a) and (8.12b) allows important considerations to be made. First, if noise is added to the amplifier's input or within the direct amplifier path, feedback does not have any effect on the equivalent input noise. Second, noise injected before or after the feedback summing node has the same effect. In contrast, the effect of noise injected at the output (for instance due to the noise of an additional load element) depends on whether or not feedback is applied. Finally, it should be pointed out that if the feedback block is implemented with noisy components, the additional source of noise, which is absent in the original open-loop amplifier, contributes to increasing the equivalent input noise. This aspect is better explained by considering the effect of feedback in an amplifier model with real feedback networks, as discussed below.

Consider the circuit in Fig. 8.8. It shows a voltage amplifier (characterised by input impedance z_{in} and by a large open-loop gain A_v) with parallel-mixing feedback at the input. Amplifier noise is modeled via generators $\overline{v}_n$ and $\overline{i}_n$, and $\overline{i}_{nR_F}$ accounts for the resistor noise.

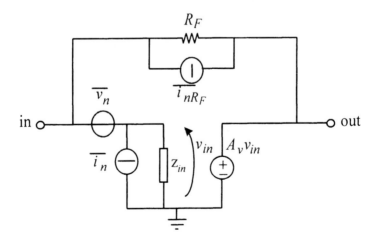

Fig. 8.8. Parallel-mixing noisy amplifier.

The equivalent noise representation of the closed-loop amplifier using the input-referred noise generators, $\overline{v_{n,in}}$ and $\overline{i_{n,in}}$, is illustrated in Fig. 8.9.

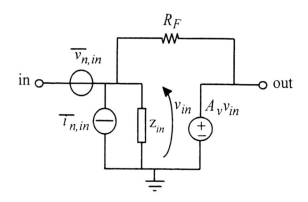

Fig. 8.9. Parallel-mixing amplifier with equivalent input noise generators.

To find the expressions of $\overline{v_{n,in}}$ and $\overline{i_{n,in}}$ (with PSD values S_{Vin} and S_{Iin}, respectively) we can follow the procedure described in Paragraph 8.2. Specifically, to calculate $\overline{v_{n,in}}$, we equate the output noise of the two circuits with their inputs shorted to ground. Under this condition, the amplifier input voltage of the circuit in Fig. 8.8 (8.9) is solely determined by $\overline{v_n}$ ($\overline{v_{n,in}}$), and is independent of the output voltage (this means that feedback is made ineffective by grounding the input). Consequently, the PSD of the output noise voltage for the two circuits is respectively

$$S_{Vout} = S_V A_v^2 \qquad (8.13a)$$

$$S_{Vout} = S_{Vin} A_v^2 \qquad (8.13b)$$

and by equating (8.13a) and (8.13b) we obtain

$$S_{Vin} = S_V \qquad (8.14)$$

We repeat the computation to evaluate $\overline{i_{n,in}}$, but now with the inputs left open. In this case feedback is effective, as the amplifier input voltage also depends on v_{out}. To simplify gain expressions, we evaluate their asymptotic values (i.e., assuming A_v to be infinitely large). In particular, the input-output transresistance closed-loop gain is R_F. The PSD of the output noise voltage for the two circuits is respectively

$$S_{Vout} = S_I R_F^2 + S_{IR_F} R_F^2 + S_V \tag{8.15a}$$

$$S_{Vout} = S_{Iin} R_F^2 \tag{8.15b}$$

Thus we get

$$S_{Iin} = S_I + S_{IR_F} + \frac{S_V}{R_F^2} \tag{8.16}$$

Equations (8.14) and (8.16) although derived for a particular circuit, are general in their essence. They show that the equivalent input noise voltage of an amplifier is unmodified by the application of parallel-mixing feedback, whilst the noise current of the feedback network is added directly to the closed-loop input-referred noise current. Moreover, when the amplifier noise voltage contributes to the equivalent closed-loop input noise current, this contribution is usually negligible. Otherwise, correlation between $\overline{v_{n,in}}$ and $\overline{i_{n,in}}$ is introduced.

Consider now the circuit in Fig. 8.10 which shows a voltage amplifier with series-mixing feedback at the input. Resistors noise is modeled via the two voltage generators $\overline{v_{nR_1}}$ and $\overline{v_{nR_2}}$.

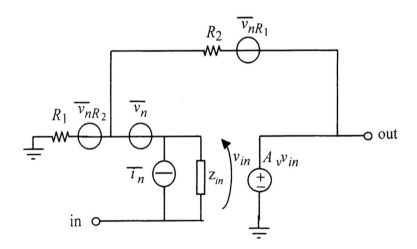

Fig. 8.10. Series-mixing noisy amplifier.

The equivalent noise representation of the closed-loop amplifier using the input-referred noise generators, $\overline{v_{n,in}}$ and $\overline{i_{n,in}}$, is illustrated in Fig. 8.11.

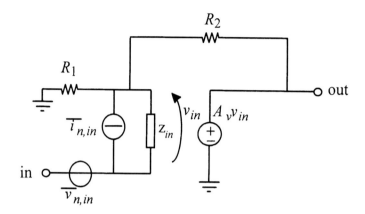

Fig. 8.11. Series-mixing amplifier with equivalent input noise generators.

Again, to calculate $\overline{v_{n,in}}$ we equate the output noise of the two circuits with the input shorted to ground. Assuming the amplifier open-loop gain to be infinitely large, the input-output closed-loop gain will be $(R_1+R_2)/R_1$, and the PSD of the output noise voltage for the two circuits will be respectively

$$S_{Vout} = S_V \left(\frac{R_1 + R_2}{R_1} \right)^2 + S_{VR_1} \left(\frac{R_2}{R_1} \right)^2 + S_{VR_2} + S_I R_2^2 \tag{8.17a}$$

$$S_{Vout} = S_{Vin} \left(\frac{R_1 + R_2}{R_1} \right)^2 \tag{8.17b}$$

from which we get

$$S_{Vin} = S_V + S_{VR_1} \left(\frac{R_2}{R_1 + R_2} \right)^2 + S_{VR_2} \left(\frac{R_1}{R_1 + R_2} \right)^2 + S_I \left(\frac{R_1 R_2}{R_1 + R_2} \right)^2 \tag{8.18}$$

$$= S_V + S_{VR_F} + S_I R_F^2$$

where in the last expression we defined $R_F = R_1 \| R_2$ and used equations (8.9a)-(8.9b).

To evaluate $\overline{i_{n,in}}$ we now leave the input open. Under this condition, the amplifier input voltage for the circuit in Fig. 8.10 (8.11) is solely determined by $\overline{i_n}$ ($\overline{i_{n,in}}$), and is independent of the output voltage. Hence, the PSD of the output noise voltage for the two circuits is respectively

$$S_{Vout} = S_I \left| A_v z_{in} \right|^2 \tag{8.19a}$$

$$S_{Vout} = S_{Iin} \left| A_v z_{in} \right|^2 \tag{8.19b}$$

and by equating the two above equations we get

$$S_{Iin} = S_I \tag{8.20}$$

Equations (8.18) and (8.20) show that the equivalent input noise current of an amplifier is unmodified by the application of series-mixing feedback, while the noise voltage of the feedback network adds directly to the closed-loop input-referred noise voltage.

Chapter 9

EXAMPLES OF FEEDBACK IN INTEGRATED CIRCUITS

In this chapter we will describe some selected examples in which the concepts developed previously are applied. These examples are select in the sense that they are related to special aspects of feedback circuits which are only marginally or never treated in standard textbooks. The aim is hence not to repeat analyses that can be found elsewhere, but to stimulate the reader with a number of case studies covering not only traditional IC analog circuits (such as the differential amplifier and current mirrors) but also analog state-of-the-art topologies (such as the current-feedback operational amplifier).

9.1 THE OUTPUT RESISTANCE OF A DIFFERENTIAL AMPLIFIER WITH CURRENT-MIRROR LOAD

The *differential amplifier* is a fundamental building block in analog integrated circuits. It is used in the implementation of the input stage of operational amplifiers and its features are assumed here to be well known to the reader. In the discussion to follow, we will consider a differential amplifier with a current-mirror active load and, specifically, we evaluate its output resistance.

The schematic diagram of the differential amplifier with current mirror active load implemented with generic transistors is illustrated in Fig. 9.1. It is made up of transistor couples T1-T2 and T3-T4, which compose the differential pair and the active load, respectively, and the constant current generator I_B. Observe that the bulk terminal has been omitted to preserve simplicity.

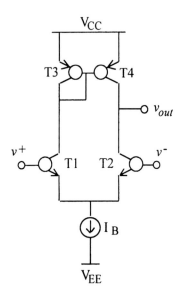

Fig 9.1. Schematic diagram of a differential amplifier with current mirror load.

The circuit exhibits a local feedback (through the current mirror T3-T4) which must be taken into account to accurately evaluate its output resistance r_o. To this end, we present the simplified AC diagram in Fig. 9.2, where the current mirror is modelled with a unitary controlled current source, i and its associated output resistance, r_{o4}, equal to the output resistance of transistor T4 (the mirror input resistance, approximately equal to $1/g_{m3}$, has been neglected as it is connected in series to the high-impedance Z terminal of transistor T1). Finally, resistor r_b models the output resistance of the current source I_B.

The output resistance is due to the parallel combination of the up and down contributions,

$$r_{out} = r_u \| r_d = r_{o4} \| r_d \qquad (9.1)$$

where r_u is exactly the same as the output resistance seen at terminal Z of transistor T4, r_{o4}.

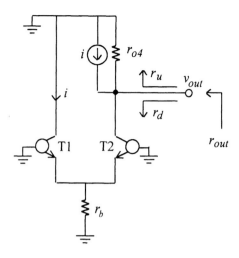

Fig 9.2. AC schematic diagram of a differential amplifier.

In contrast to what a superficial analysis could show, the down resistance is not simply equal to the resistance seen by looking into terminal Z of transistor T2. The down resistance is affected by the feedback provided by the current mirror which injects almost the same current flowing through resistance r_{z1} into the output node. Hence, a unitary loop gain ($T = 1$) is experienced by the down resistance which is reduced by a factor of two compared to its open-loop value. We can calculate this open-loop value by setting the controlled current generator to zero, and recognising that transistor T2 is in common Y configuration with an equivalent degenerative resistance R_{X2}, given by

$$R_{X2} = \frac{r_{p1} \| r_b}{1 + g_{m1} r_{p1}} \approx \frac{r_{p1}}{1 + g_{m1} r_{p1}} \tag{9.2}$$

which is much smaller than r_{p1}.
    Thus, the open-loop resistance is

$$r_{z2} = r_{o2} + (1 + g_{m2} r_{o2}) r_{p2} \| R_{X2} \tag{9.3}$$

Substituting (9.2) into (9.3) we obtain

$$r_{z2} = 2r_{o1} \tag{9.4}$$

and the close-loop value is

$$r_d = \frac{r_{z1}}{1+T} = \frac{r_{z1}}{2} = r_{o1} \tag{9.5}$$

Finally, we find the well-known result for the output resistance of a differential stage [GM93], [LS94], [JM97]

$$r_{out} = r_{o4} \| r_{o1} \tag{9.6}$$

9.2 THE WILSON CURRENT MIRROR

Among the several current mirror topologies, the Wilson current mirror is a high-performance solution which is heavily based on a negative-feedback configuration, [W90], [W901]. The AC schematic diagram of the Wilson current mirror is depicted in Fig. 9.3 (again, bulk connections are omitted for simplicity). Note that although the circuit is described using the generic transistor, in common practice it is more frequently encountered in its bipolar implementation. In fact, as transistors T1 and T2 work with substantially different Z-to-X voltages, this can cause offset and gain error especially with MOS processes in which the channel length modulation is more significant than the corresponding Early effect of bipolar transistors.

The feedback mechanism can simply be explained as follows: the output current, i_{out}, is collected into the input of current mirror T2-T1, and is then fed back to the mirror input so it can be subtracted from the input signal, i_s (negative shunt-series feedback)

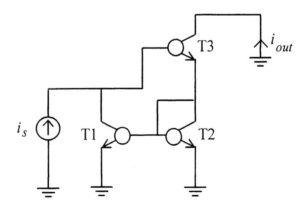

Fig 9.3. AC schematic diagram of the Wilson current mirror.

To simplify the analysis, consider the mixed model of the circuit under consideration shown in Fig. 9.4, where small-signal models of transistors T1 and T2 are drawn. Note also that the diode-connected transistor T2 is equivalent to the parallel of resistance $1/g_{m2}$ and r_{p2}. Since resistance $1/g_{m2}$ is always much lower than both resistances r_{p1} and r_{p2}, the mixed model can be further simplified as shown in Fig. 9.5.

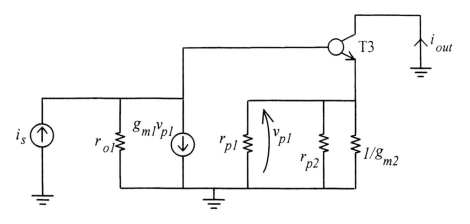

Fig 9.4. Mixed model of the Wilson current mirror.

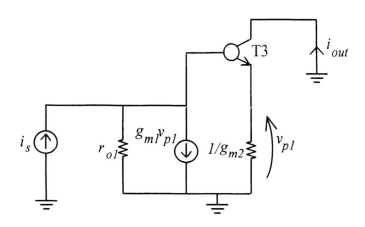

Fig 9.5. Simplified mixed model of the Wilson current mirror.

Now we use the Rosenstark/Blackman equations to find the asymptotic gain and the input and output resistances, assuming that the controlled generator characterised by transconductance g_{m1} is the critical parameter. Of course, setting this parameter to zero means switching off transistor T1 allowing the direct transmission term to be evaluated from Fig. 9.6. The output current can be simply calculated by considering the equivalent transconductance, g_{meq3} of transistor T3 in a common X configuration with a degenerative resistance equal to $1/g_{m2}$. Assuming all the transistors with equal small-signal parameters, this equivalent transconductance is about equal to $g_m/2$ (see (2.15b)), and G_o is

$$G_o = \left.\frac{i_{out}}{i_s}\right|_{g_{m1}=0} = g_{meq3}r_{o1}\|r_{y3} \approx \frac{g_m}{2}r_o\|2r_p \tag{9.7}$$

which is greater than unity.

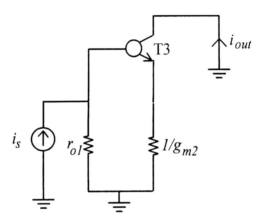

Fig 9.6. Model to evaluate the direct transmission term of the Wilson current mirror.

To evaluate the return ratio we set i_s to zero and replace the original controlled current generator, $g_{m1}v_{p1}$, with an independent current source, i, as shown in Fig 9.7. Aside from the opposite flow direction of the independent current generators, Fig. 9.5 and 9.6 are identical. Therefore, the return ratio exactly equals the direct transmission term (i.e., $T = G_o$).

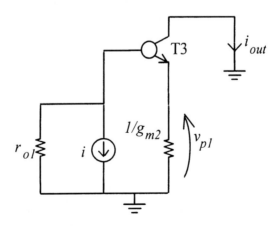

Fig 9.7. Model to evaluate the return ratio of the Wilson current mirror.

Now we set parameter g_{m1} infinitely large to calculate the closed-loop asymptotic gain. Returning to Fig. 9.5, this means that voltage v_{p1} must be zero, as this condition holds only if the output current is zero. Thus $G_\infty = 0$ and, as expected in a current mirror, the closed loop transfer function (current gain) is very close to one

$$G_F = \frac{i_{out}}{i_s} = \frac{G_o}{1+T} = \frac{T}{1+T} \qquad (9.8)$$

The driving point input and output resistances, $r_{in,ol}$, and $r_{out,ol}$, are calculated using the results in Chapter 2. They are given by

$$r_{in,ol} = r_o \| 2r_p \qquad (9.9)$$

$$r_{out,ol} = (1 + \lambda_b)r_o + r_p \approx r_o \qquad (9.10)$$

Finally we evaluate $T(0, R_L)$, $T(\infty, R_L)$, $T(R_S, 0)$ and $T(R_S, \infty)$ to find the input and output resistances

$$T(0, R_L) = 0 \qquad (9.11a)$$

$$T(\infty, R_L) = T \qquad (9.11b)$$

$$T(R_S, 0) = T \qquad (9.11c)$$

$$T(R_S, \infty) = 1 \qquad (9.11d)$$

with the input and output resistances equal to

$$r_{in} = \frac{r_o \| 2r_p}{1+T} \approx \frac{2}{g_m} \qquad (9.12)$$

$$r_{out} = r_o \frac{1+T}{2} \approx \frac{g_m}{4}(r_o \| 2r_p) r_o \qquad (9.13)$$

In conclusion, for a bipolar Wilson current mirror we get

$$G_F = \frac{i_{out}}{i_s} = \frac{1}{1+1/\beta} \qquad (9.14)$$

$$r_{in} \approx \frac{2}{g_m} \qquad (9.15)$$

and

$$r_{out} \approx \frac{\beta}{2} r_o \qquad (9.16)$$

9.3 THE CASCODE CURRENT MIRROR

The cascode current mirror avoids the drawbacks of the Wilson mirror by realising a symmetrical topology which sets the Z-to-X voltages of transistors T1 and T2 almost equal, thereby minimising errors due to the finite transistor output resistance [AH87], [GT86]. The AC schematic of the cascode current mirror is shown in Fig. 9.8. Note that in this case, the circuit behaviour is not based on feedback. Rather, its improved performance is achieved thanks to the current-buffering action of the cascode transistor T4. In this manner, the accuracy of the current gain and the output resistance are increased. In the following we will not analyse these characteristics, as they seem self-evident. Instead, we will concentrate our attention on a special effect due to a local feedback that arises only in a bipolar implementation. As we will see, this effect slightly reduces the achievable output resistance.

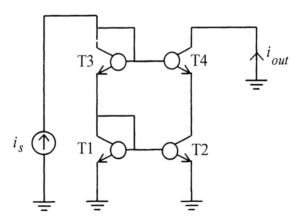

Fig 9.8. AC schematic diagram of the cascode current mirror.

From the AC schematic diagram we derive the small-signal circuit of the cascode current mirror shown in Fig. 9.9.

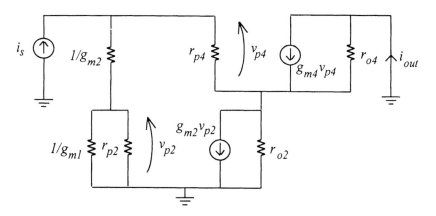

Fig 9.9. Small-signal circuit of the cascode current mirror.

The output resistance of the cascode current mirror is that of the common Y transistor T4, which is almost equal to

$$r_{out} = r_{z4} = r_{o4} + \left(1 + g_{m4}r_{o4}\right)R_{X4} \tag{9.17}$$

By superficial analysis of the circuit in Fig. 9.9 we could erroneously assume R_{X4} to be about equal to the parallel between r_{o2} and r_{p4} (where the series contribution to r_{p4}, given by $1/g_{m1} + 1/g_{m2}$ has been neglected). However, if resistance r_{p4} is finite, there is a feedback loop that reduces this value. Indeed, by choosing transconductance g_{m2} as a critical parameter, we find that term $T(R_S, 0)$ is much lower than one and can be neglected. Otherwise, $T(R_S, \infty)$ is almost equal to one. Hence, resistance R_{X4} is equal to its value assuming transconductance g_{m2} to be equal to zero divided by two

$$R_{X4} \approx \frac{r_{o2} \| r_{p4}}{1 + T(R_S, \infty)} \approx \frac{r_{p4}}{2} \tag{9.18}$$

9.4 THE CURRENT FEEDBACK OPERATIONAL AMPLIFIER AND ITS HIGH-LEVEL CHARACTERISTICS

Recently, we have witnessed the affirmation of a novel op-amp architecture now available from several of the specialist analogue semiconductor manufacturers. These op-amps are generally referred to as Current-Feedback Operational Amplifiers (CFOAs) [S911], [TLH90],

[G93], [B93], [SKW94], and represent an evolution in the architecture of the voltage-mode op-amps (VOAs), which have otherwise remained much the same over the years. Implementations of high-performance CFOAs have become possible thanks to the availability of high quality complementary bipolar transistors provided by advanced BJT processes. However, the low transconductance of MOS transistors, does not make this component suitable for implementing CFOAs. The internal architecture of a CFOA is exemplified in Fig. 9.10. It is made up of a CCII-based input stage –that performs a voltage following action between terminal Y to X and a current following action between X and Z– and an output stage with a voltage buffering function which properly drives the load and isolates it from the internal high-resistance, R_t, at node Z. The amplifier is hence characterised by an inverting low-impedance terminal and by a transresistance gain v_{out}/i^- equal to R_t (this is the reason for the name "transimpedance amplifier" also used when referring to this opamp).

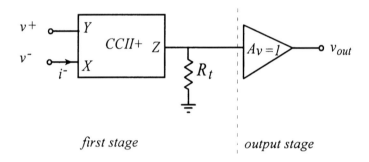

first stage   *output stage*

Fig 9.10. Typical CFOA architecture.

The CFOA circuit symbol is not different from that of a conventional single-ended VOA, as can be deduced from Fig. 9.11. The same figure also explicitly shows in a dotted line, the input voltage follower and the output current-controlled voltage source. Observe that the voltage follower will operate outside the feedback loop (which involves the output and the inverting input terminal, but not the noninverting input). This fact can result in a source of harmonic distortion especially in noninverting configurations in which the CFOA is operated under large common mode signal swings. Nevertheless, it must be observed that although the internal structure of the CFOA differs greatly from that of a traditional VOA, the external feedback circuitry and its applications are similar to those of a VOA [S96], [S911] [SKW94], [TL94]. Hence, a variety of configurations with their respective performance parameters have been directly derived from traditional ones, without requiring any further study.

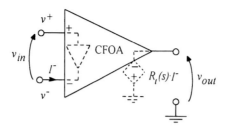

Fig 9.11. CFOA circuit symbol with an input voltage follower and an output current controlled voltage source.

CFOAs have become popular because of their inherent advantages: excellent small-signal and large signal performance and, under proper operating conditions, a closed-loop bandwidth independent of gain. In fact, thanks to the low-impedance input terminal, the typical closed-loop bandwidth is in the range of 50 to 100 MHz and a very high slew rate capability in the range of 500- to 5000-V/μs [B90], [S911], [SS92], [G93], [B93], [SKW94], [B97] has been reported. The high slew rate performance derives from the use of a class AB topology for both the input and output stages. Although the differential stages with class AB capability, reported in the literature [SW87], [BW83], [C93], [CMPP93], could be used in conventional opamps to provide a high slew-rate, this option is rather unusual and expensive.

The closed-loop constant-bandwidth property offered by a CFOA can be easily explained by considering the closed-loop configuration shown in Fig. 9.12.

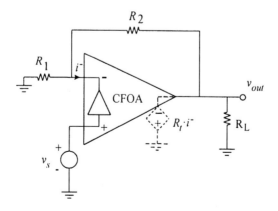

Fig.9.12. CFOA in non-inverting feedback configuration.

The asymptotic gain and the loop gain of the circuit can be found using the Rosentark method, choosing the dependent output voltage generator as critical source. Since R_t is made infinitely large, current i tends to zero. Thus the same current flows into R_1 and R_2 and since voltage v_s appears at the inverting terminal (thanks to the input voltage follower) a virtual short-circuit appears between the inverting and the non inverting terminals. The asymptotic closed loop gain is hence

$$G_\infty = 1 + \frac{R_2}{R_1} \tag{9.19}$$

which is the same result we would have obtained using a VOA in place of the CFOA.

In addition, the loop gain is

$$T = \frac{R_t}{R_2} \tag{9.20}$$

showing that the loop gain depends on the transresistance gain, R_t, and on only one of the two external resistances, R_2. Since the closed-loop bandwidth is proportional to the loop gain, and the closed-loop gain can be set by changing only R_1, a closed-loop constant bandwidth behavior is achieved.

9.5 TRANSISTOR-LEVEL ARCHITECTURE, SMALL-SIGNAL MODEL, AND FREQUENCY COMPENSATION OF CFOAs

According to the architecture in Fig. 9.9, a CFOA is made up of three main blocks: two voltage buffers and one current buffer. More specifically, the first voltage buffer is located at the input and its output current (I^-) is replicated by the current buffer into a high-impedance internal node. The other voltage buffer is at the output with the purpose of properly driving the output load.

The simplified circuit schematic of a typical CFOA is represented in Fig. 9.13, where bipolar transistors are used [HR80], [TLH90]. The Load and compensation capacitors, C_L and C_t, and the feedback network made up of resistors R_1 and R_2, are also included. The input voltage buffer is implemented with transistors T1-T4 and associated bias current generators I_{B1}. Two current mirrors T5-T6 and T7-T8 implement the current buffer, while transistors T9-T11 and generators I_{B2} form the output voltage buffer. Capacitor C_t provides dominant-pole compensation. Nearly all monolithic complementary bipolar high-speed CFOAs are a variation of this architecture [B97].

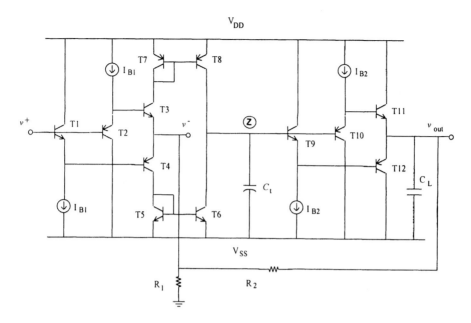

Fig. 9.13. CFOA schematic diagram with feedback resistors and capacitive load.

It is worth noting that the input voltage buffer is outside the feedback loop and hence does not affect the frequency response and stability of the amplifier. The feedback loop of the amplifier includes the output voltage buffer (T9-T12), resistances R_1 and R_2, transistors T3 and T4, which work as simple common-base amplifiers, and the two current mirrors T5-T6 and T7-T8.

A simple small-signal model of the CFOA is shown in Fig. 9.14 where R_t is the equivalent resistance at gain node Z. The output resistance of the input voltage buffer, $1/g_{mi}$, is the input resistance at inverting node, and the input resistance of the input buffer, r_{bi}, is the input resistance of the noninverting node. The output resistance of the output voltage buffer, $1/g_{mo}$, is the output resistance. Controlled generators α_1, α_2, and h model the transfer functions of the input and output voltage buffer, and the complementary current mirror, respectively, and are usually almost unitary in module.

By inspection of Fig. 9.13, we get

$$R_t \approx \frac{\beta_n}{2} r_{z6} \left\| \frac{\beta_p}{2} r_{z8} \right. \tag{9.21}$$

$$\frac{1}{g_{mi}} = \frac{1}{g_{m3} + g_{m4}} \tag{9.22}$$

$$\frac{1}{g_{mo}} = \frac{1}{g_{ml1} + g_{ml2}} \tag{9.23}$$

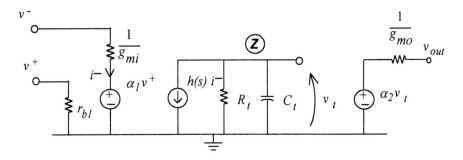

Fig. 9.14. CFOA small signal model.

Stability and related frequency compensation issues of CFOAs have been treated in [MT96], but the approach adopted is not simple from a design point of view and leads to unnecessary conditions being placed on the feedback network. Moreover, this analysis neglects a fundamental contribution that arises when the CFOA load is capacitive, which is an usual circumstance.

In order to apply the compensation procedures described in Chapters 4 and 5, we need to evaluate the return ratio. After substituting the CFOA small signal model in the circuit in Fig. 9.13 we choose the voltage gain of buffer α_2 as the critical parameter. Setting the input source to zero and representing the critical controlled voltage source with v' we get the equivalent circuit in Fig. 9.15.

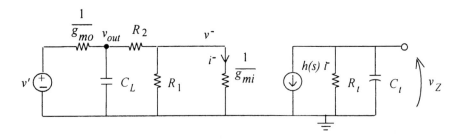

Fig. 9.15. Small-signal circuit of CFOA with feedback resistances to evaluate the return ratio.

The DC open loop gain results as

$$A_{ol}(0) = \frac{v_Z}{v'} = \frac{R_1}{\dfrac{1}{g_{mi}} + R_1} \frac{R_t}{\dfrac{1}{g_{mo}} + R_2 + R_1 \left\| \dfrac{1}{g_{mi}} \right.} =$$

$$= g_{mi} R_1 \frac{R_t}{\left(g_{mi} R_1 + 1\right)\left(R_2 + \dfrac{1}{g_{mo}}\right) + R_1}$$

(9.24a)

and if resistance R_1 is much greater than $1/g_{mi}$, it simplifies to

$$A_{ol}(0) \approx \frac{R_t}{\dfrac{1}{g_{mo}} + R_2 + \dfrac{1}{g_{mi}}}$$

(9.24b)

This shows that unlike for voltage-mode operational amplifiers, resistance R_1, if properly dimensioned, must not be considered as a component of the feedback network.

The dominant pole is created at the high resistance node, Z. It is simply expressed by

$$p_1 = \frac{1}{R_t C_t}$$

(9.25)

and the gain-bandwidth product is

$$\omega_{GBW} = \frac{1}{\left(R_2 + \dfrac{1}{g_{mo}} + \dfrac{1}{g_{mi}}\right) C_t}$$

(9.26)

Equation (9.26) shows the well-known property of CFOAs: the closed-loop bandwidth is independent of the closed-loop gain provided that resistance R_2 is maintained constant. It is apparent that we have the highest gain-bandwidth product when $R_2 = 0$, which means the CFOA is in unity-gain configuration. Thus, in order to guarantee stability with safe margins, we have to compensate the CFOA in the unity-gain configuration as for the traditional voltage operational amplifier. Moreover, compensation can be achieved by adopting the dominant pole approach, which requires a suitable increase in capacitance C_t at the high-resistance node, Z. In order to provide a given phase margin, ϕ, the relationship

$$\omega_{GBW} = p_2 / \tan(\phi) \qquad (9.27)$$

holds, where p_2 is the equivalent second pole generally due to load capacitance C_L [P99]

$$p_2 = \cfrac{1}{\left(R_2 + \cfrac{1}{g_{mi}} \Vert R_1\right) \left\Vert \cfrac{1}{g_{mo}} \right. C_L} \qquad (9.28)$$

Even if there is no load capacitor outside the chip, there is a contribution internal to the IC caused by the bounding pad and pin capacitances.

Other frequency limitations are due to the transfer functions $\alpha_2(s)$ and $h(s)$ characterised by high-frequency poles given that they arise at low resistance nodes. They can therefore be neglected in a first-order analysis. Observe, however, that those poles associated with the output voltage buffer transfer function are much higher than those of the current mirrors.

In conclusion, using (9.26), with R_2 set to zero, in relationship (9.27) we get

$$C_t = \tan(\phi)\frac{g_{mi}g_{mo}}{\left(g_{mi} + g_{mo}\right)^2} \qquad (9.29)$$

9.6 INTEGRATORS AND DIFFERENTIATORS WITH CFOAs

As a useful application we shall now consider the implementation of a simple integrator and differentiator using the CFOA. These circuits are shown in Figs. 9.16 and 9.17, respectively.

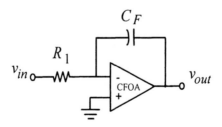

Fig. 9.16. Integrator with CFOA.

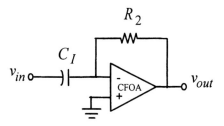

Fig. 9.17. Differentiator with CFOA.

The loop-gain transfer function of the integrator in Fig. 9.16, T_{INT}, can be obtained from that of a CFOA with a resistive feedback as calculated in the previous section, by replacing R_2 with $1/sC_F$.

$$T_{INT}(s) = \frac{R_t C_F s}{\left[1 + \left(\dfrac{1}{g_{mo}} + \dfrac{1}{g_{mi}}\right)C_F s\right]\left(1 + \dfrac{s}{p_1}\right)\left(1 + \dfrac{s}{p_2}\right)} \qquad (9.30a)$$

At frequencies higher than the two dominant poles, remembering the expression of p_1, the transfer function (9.30a) can be simplified into

$$T_{INT}(s) \approx \frac{1}{\left(\dfrac{1}{g_{mo}} + \dfrac{1}{g_{mi}}\right)C_t} \frac{1}{\left(1 + \dfrac{s}{p_2}\right)s} \qquad (9.30b)$$

Assuming a safely compensated CFOA with pole p_2 higher than the transition frequency, we get the same gain-bandwidth product given by (9.26) setting $R_2 = 0$. As a consequence, the compensation procedure is equal to the one already studied for a purely resistive feedback. This is simple to understand: for high frequencies such as those in the vicinity of the transition frequency we can assume the feedback capacitance, C_F, to be short-circuited, and the CFOA in unity gain configuration.

To evaluate the loop-gain transfer function of the differentiator in Fig. 9.17, we have to substitute resistor R_1 with capacitor C_I in the CFOA with resistive feedback. In an ideal CFOA, whose resistance at the inverting input is also nominally equal to zero, capacitance C_I would be outside the loop gain, and would not play any part. However, real CFOAs have a finite buffer output resistance. Moreover, capacitance C_I determines another high-frequency pole in the loop-gain transfer function. It can be shown that the loop-gain transfer function becomes

$$T_{DER}(s) = \frac{R_t}{R_2} \frac{1}{\left(1 + \frac{C_1}{g_{mi}}s\right)\left(1 + \frac{s}{p_1}\right)\left(1 + \frac{s}{p_2}\right)} \tag{9.31}$$

and the gain-bandwidth product is

$$\omega_{GBW,DER} = \frac{1}{R_2 C_t} \tag{9.32}$$

Hence the compensation task requires that

$$C_t \approx \frac{\tan(\phi)}{R_2}\left(\frac{C_1}{g_{mi}} + \frac{C_L}{g_{mo} + g_{mi}}\right) = \frac{\tan(\phi)}{g_{mi}R_2}\left(C_1 + \frac{C_L}{2}\right) \tag{9.33}$$

It is worth noting that we can compensate the CFOA with a capacitance much lower than C_1. This is a very different condition from that required in a traditional opamp, where a high capacitance value, often impractical with IC technologies, is needed to compensate the differentiator configuration [WHK92], [A88], [P99].

9.7 CFOA VERSUS VOA

In this Section we make a brief comparison between the bipolar CFOA and VOA in regard to static and frequency response performance [PP01]. The comparison assumes that actual CFOA behavior is characterized by a dominant pole and a second equivalent pole, limiting the amplifier gain-bandwidth product. The comparison is with a VOA of comparable topology, thus providing similar features. The same power consumption is assumed for both amplifiers.

The VOA topology chosen is the *folded cascode* one shown in Fig. 9.18.

The main characteristic of this topology is, like the CFOA in Fig. 9.13, having only one high-gain stage, since it achieves the high voltage gain thanks to the high equivalent resistance at node A. Moreover, the full transconductance of the input differential stage is gained by using the Wilson current mirror T4-T6, that performs a differential-to-single conversion.

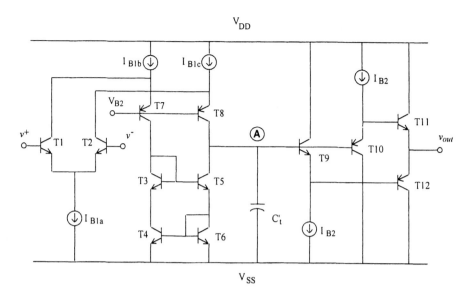

Fig. 9.18. Schematic diagram of the folded-cascode VOA.

The equivalent small-signal model of the VOA considered is shown in Fig. 9.19, where R'_t and C'_t are the equivalent resistance and the compensation capacitance at the gain node, respectively, and $2r_p$ is the equivalent resistance at the input of the differential stage T1-T2. The transconductance gain, g_{m1}, is equal to that of the input transistors T1-T2, with the other parameters being previously defined.

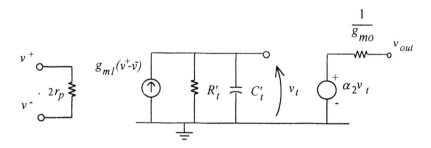

Fig. 9.19. Small signal model of the VOA.

Applying a resistive feedback as shown in Fig. 9.12, the DC loop gain becomes

$$T(0) = 2\frac{2r_p \| R_1}{R_2 + \dfrac{1}{g_{mo}} + 2r_p \| R_1} g_{m1} R'_t \approx 2\frac{g_{m1}}{G_\infty} \qquad (9.34)$$

where G_∞ is defined in (9.19).

The dominant pole of the open-loop amplifier is again given by (9.25) after substituting R'_t and C'_t for R_t and C_t. Thus, the gain-bandwidth product results

$$\omega_{GBW} \approx 2\frac{g_{m1}}{G_\infty C'_t} \tag{9.35}$$

For VOA, too, the second pole of the loop-gain can be assumed at the output and is given by

$$p_2 = \frac{1}{(R_1 + R_2)\left\|\frac{1}{g_{mo}}\right\| C_L} \approx \frac{g_{mo}}{C_L} \tag{9.36}$$

Hence, to achieve phase margin ϕ under the worst condition of unity-gain loop, the compensation capacitance

$$C'_t = tg(\phi)\frac{2g_{m1}}{g_{mo}}C_L \tag{9.37}$$

is needed.

Since a trade-off exists between frequency performance and power dissipation (and sometimes between gain and power dissipation), we assume, without loss of generality, the same power consumption for both the CFOA and VOA by setting $I_{B1a,b,c}$ (of VOA in Fig. 9.18) equal to $2I_{B1}$ (of CFOA in Fig. 913). Consequently, the transconductance of the VOA input stage results equal to the input resistance at the inverting terminal of the CFOA

$$g_{m1} = g_{mi} = g_{mo} = g_m \tag{9.38}$$

Moreover, the VOA resistance R'_t is in the range $R_t < R'_t < 2R_t$ (it is equal to $2R_t$ if a cascode current mirror is used in the VOA instead of a Wilson current mirror).

Comparing the open-loop gain of the CFOA with that of the VOA, we get

$$\frac{A_C(0)}{A_V(0)} = \frac{R_t}{R'_t}\frac{G_\infty}{2g_m R_2 + 2(1 + G_\infty)} \tag{9.39}$$

which is always much lower than 1. It is reduced by decreasing the closed-loop gain and tends to $R_t/4R'_t$ when the amplifiers are used in unity-gain configuration. This means that for the same amount of power consumption, the accuracy of a bipolar CFOA is worse than that achieved with a VOA. However, if we compare the resulting bandwidth of the CFOA and VOA we get

$$\frac{\omega_{dC}}{\omega_{dV}} = \frac{R'_t C'_t}{R_t C_t} = 8 \frac{R'_t}{R_t} \tag{9.40}$$

which shows that the CFOA is superior to the VOA for the same power consumption by about one order of magnitude. This advantage in terms of speed is achieved in an open-loop configuration. But to really evaluate the speed benefit we have to compare the frequency response in a closed-loop configuration. Since the closed-loop bandwidth is equal to the gain-bandwidth product of the open-loop gain, we can simply compare the gain-bandwidth product of the two amplifiers

$$\frac{\omega_{GBW,C}}{\omega_{GBW,V}} = 4 \frac{G_\infty}{g_m R_2 + 1 + G_\infty} \tag{9.44}$$

The above equation reveals that when the amplifiers are used in unity-gain configuration (with R_2 equal to zero), the CFOA gain-bandwidth product is only twice as great as that of the VOA considered. Moreover, the bandwidth improvement is not so marked for closed-loop gains greater than 1, since term $g_m R_2$ must be greater than 1 (the CFOA gain-bandwidth is greater than the VOA's only if condition $g_m R_2 < 3G_\infty - 1$ is met).

Appendix

FREQUENCY ANALYSIS OF RC NETWORKS

In this book we have studied numerous configurations whose small-signal equivalent circuits are made up of only three components, namely resistors, capacitors, and controlled sources. The evaluation of fundamental circuit parameters such as voltage and current gain transfer functions, but also input and output impedances, is then placed in the context of the study of generalised RC networks.

A.1 TRANSFER FUNCTION OF A GENERIC RC NETWORK

Consider the RC network in Fig. A.1, which includes n independent capacitors. The transfer function between two generic network ports, as a function of the complex frequency s, can be expressed in the general form

$$G(s) = G_o \frac{1 + b_1 s + b_2 s^2 + ... + b_m s^m}{1 + a_1 s + a_2 s^2 + ... + a_n s^n} \tag{A.1}$$

where $m < n$. Transfer function $G(s)$ can be determined by applying Kirchhoff's laws to the network. But when the order, n, is high, or in other words, when there is a large number of reactive components, we have to solve a system with a great number of equations. Because of this, it would be more practical to have a systematic method by which coefficients a_i and b_i could be directly determined. This approach would also have the advantage of allowing an approximate transfer function of reduced order, defined by the user, to be derived.

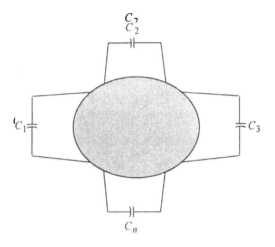

Fig. A.1. Generic RC network.

Coefficients a_i can be evaluated by using the time constant method demonstrated in [CG73]

$$a_1 = \sum_{i_1=1}^{n} \alpha_{i_1} \tag{A.2a}$$

$$a_2 = \sum_{i_1=1}^{n-1} \sum_{i_2=i_1+1}^{n} \alpha_{i_1 i_2} \tag{A.2b}$$

$$a_3 = \sum_{i_1=1}^{n-2} \sum_{i_2=i_1+1}^{n-1} \sum_{i_3=i_2+1}^{n} \alpha_{i_1 i_2 i_3} \tag{A.2c}$$

$$a_r = \sum_{i_1=1}^{n-r+1} \sum_{i_2=i_1+1}^{n-r+2} \cdots \sum_{i_r=i_{r-1}+1}^{n} \alpha_{i_1 i_2 \ldots i_r} \tag{A.2d}$$

where

$$\alpha_{i_1} = R_{i_1} C_{i_1} \tag{A.3a}$$

$$\alpha_{i_1 i_2} = R_{i_1} R_{i_2}^{i_1} C_{i_1} C_{i_2} \tag{A.3b}$$

$$\alpha_{i_1 i_2 i_3} = R_{i_1} R_{i_2}^{i_1} R_{i_3}^{i_1 i_2} C_{i_1} C_{i_2} C_{i_3} \tag{A.3c}$$

$$\alpha_{i_1 i_2 \ldots i_r} = R_{i_1} R_{i_2}^{i_1} \ldots R_{i_r}^{i_1 i_2 \ldots i_r} C_{i_1} C_{i_2} \ldots C_{i_r} \qquad \text{(A.3d)}$$

In relationships (A.3), R_j is the equivalent resistance seen from capacitor C_j whilst the other capacitors are assumed to be open (i.e., with a zero value). The equivalent resistance $R_j^{i_1 i_2 \ldots i_r}$ is the one seen from capacitor C_j when capacitors $C_{i_1} \ldots C_{i_r}$ are short circuited (i.e., with an infinite value) and the others are assumed to be open.

Coefficients b_i are evaluated by using the methods in [DM80] which give

$$b_1 = \frac{\displaystyle\sum_{i_1=1}^{n} H_{i_1} \alpha_{i_1}}{G_o} \qquad \text{(A.4a)}$$

$$b_2 = \frac{\displaystyle\sum_{i_1=1}^{n-1} \sum_{i_2=i_1+1}^{n} H_{i_1 i_2} \alpha_{i_1 i_2}}{G_o} \qquad \text{(A.4b)}$$

$$b_3 = \frac{\displaystyle\sum_{i_1=1}^{n-2} \sum_{i_2=i_1+1}^{n-1} \sum_{i_3=i_2+1}^{n} H_{i_1 i_2 i_3} \alpha_{i_1 i_2 i_3}}{G_o} \qquad \text{(A.4c)}$$

$$b_r = \frac{\displaystyle\sum_{i_1=1}^{n-r+1} \sum_{i_2=i_1+1}^{n-r+2} \ldots \sum_{i_r=i_{r-1}+1}^{n} H_{i_1 i_2 \ldots i_r} \alpha_{i_1 i_2 \ldots i_r}}{G_o} \qquad \text{(A.4d)}$$

where coefficients α are those in (A.3) and parameters $H_{i_1 i_2 \ldots i_r}$ are the DC transfer functions between the input and the output when capacitors $C_{i_1} \ldots C_{i_r}$ are short circuited.

In order to exemplify the procedure, consider the RC network in Fig. A.2 which includes three capacitors. The goal is to find its transfer function in the same form as expressed in (A.1)

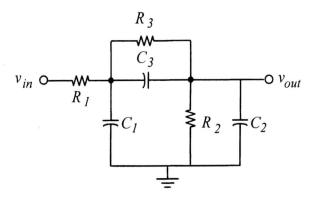

Fig. A.2. RC network used in the example.

The DC gain is found by opening all the capacitors. It is equal to

$$G_o = \frac{R_2}{R_1 + R_2 + R_3} \tag{A.5}$$

Coefficients a_i and b_i are found using (A.2)-(A.4)

$$a_1 = R_1 \| (R_2 + R_3)C_1 + R_2 \| (R_1 + R_3)C_2 + R_3 \| (R_1 + R_2)C_3 \tag{A.6a}$$

$$a_2 = \left[R_1 \| (R_2 + R_3)\right]\!\left(R_2 \| R_3\right)C_1C_2 + \left[R_1 \| (R_2 + R_3)\right]\!\left(R_2 \| R_3\right)C_1C_3 + \tag{A.6b}$$

$$+ \left[R_2 \| (R_1 + R_3)\right]\!\left(R_1 \| R_3\right)C_2C_3$$

$$a_3 = 0 \tag{A.6c}$$

$$b_1 = \frac{\dfrac{R_2}{R_1 + R_2} R_3 \| (R_1 + R_2)C_3}{G_o} = R_3C_3 \tag{A.6d}$$

$$b_2 = 0 \tag{A.6e}$$

The order of the transfer function is 2 because only two of the capacitors in Fig. A.2 are independent (indeed, C_1, C_2 and C_3 form a loop of capacitors).

A.2 APPROXIMATED POLES

In many practical circumstances it is useful to approximate the complete transfer function with a second-order one

$$G(s) = G_o \frac{1 + b_1 s}{1 + a_1 s + a_2 s^2} \tag{A.6}$$

and when the two poles are real, which holds for $a_2 < a_1^2/4$, we can rewrite relationship (A.6) as shown below

$$G(s) = G_o \frac{1 + \dfrac{s}{z}}{\left(1 + \dfrac{s}{p_1}\right)\left(1 + \dfrac{s}{p_2}\right)} \tag{A.7}$$

where $z = 1/b_1$ and the poles are related to coefficients a_i through [MG87] [GT86],

$$a_1 = \frac{1}{p_1} + \frac{1}{p_2} \tag{A.8a}$$

$$a_2 = \frac{1}{p_1 p_2} \tag{A.8b}$$

By adding the condition of dominant-pole behaviour, which means $p_1 \ll p_2$, relationships (A.8) lead to a simple and useful relation between poles and coefficients a_i.

$$p_1 \approx \frac{1}{a_1} \tag{A.9a}$$

$$p_2 \approx \frac{a_1}{a_2} \tag{A.9b}$$

The approach can be extended to higher-order functions with real and well spaced poles [MG87].

Recently, an interesting approximation of the roots of a second-order equation, that can be profitably used to find poles and zeros of transfer functions, has been proposed [R02].

The approximations of equation $1 + a_1 s + a_2 s^2$ are derived in ranges of parameter x defined as

$$x = \frac{4a_2}{a_1^2} \qquad\qquad (A.10)$$

and hence are different for different values of x. The roots summarised in Table A.1, show an error always lower than 10%.

Table A.1

X	roots
$X < -15$	$-0.5\dfrac{a_1}{a_2} \pm \dfrac{1}{\sqrt{|a_2|}} \qquad$ for $a_1 < 0$ $-0.5\dfrac{a_1}{a_2} \mp \dfrac{1}{\sqrt{|a_2|}} \qquad$ for $a_1 > 0$
$-15 \le x < -4$	$0.35\dfrac{a_1}{a_2} - 0.32\dfrac{1}{a_1}$ and $-1.4\dfrac{a_1}{a_2} + 0.3\dfrac{1}{a_1}$
$-4 \le x < -0.3$	$0.028\dfrac{a_1}{a_2} - 0.64\dfrac{1}{a_1}$ and $-0.95\dfrac{a_1}{a_2} + 0.76\dfrac{1}{a_1}$
$-0.3 \le x < 0.3$	$-\dfrac{1}{a_1}$ and $-\dfrac{a_1}{a_2}$
$0.3 \le x < 0.9$	$0.045\dfrac{a_1}{a_2} - 1.6\dfrac{1}{a_1}$ and $-1.15\dfrac{a_1}{a_2} + 2.2\dfrac{1}{a_1}$
$0.9 \le x < 1$	$1.35\dfrac{a_1}{a_2} - 7.4\dfrac{1}{a_1}$ and $-2.2\dfrac{a_1}{a_2} + 6.8\dfrac{1}{a_1}$
$1 \le x < 1.1$	$-(0.5 \pm 2j)\dfrac{a_1}{a_2} \pm \dfrac{8j}{a_1}$
$1.1 \le x < 4$	$-(0.5 \pm 0.085j)\dfrac{a_1}{a_2} \pm 1.04\dfrac{j}{a_1}$

$4 \leq x < 15$	$-\left(0.5 \pm 0.6j\right)\dfrac{a_1}{a_2} \pm 0.34\dfrac{j}{a_1}$
$x < 15$	$-0.5\dfrac{a_1}{a_2} \mp \dfrac{1}{\sqrt{a_2}}$ for $a_l < 0$ $-0.5\dfrac{a_1}{a_2} \pm \dfrac{1}{\sqrt{a_2}}$ for $a_l > 0$

REFERENCES

[A83]   B. Ahuja, "An Improved Frequency Compensation Technique for CMOS Operational Amplifiers," *IEEE J. of Solid-State Circuits*, Vol. SC-18, No. 6, pp. 629-633, Dec. 1983.

[A88]   M. A. Al-Alaoui, "A Stable Differentiator with a Controllable Signal-to-Noise Ratio," *IEEE Trans. on Instr. and Measur.*, Vol. 37, No. 3, pp. 383-388, Sept. 1988.

[A95]   A. Arbel, "Comparison Between the Noise Performance of Current-Mode and Voltage-Mode Amplifiers," *Analog Integrated Circuits and Signal Processing*, No.7, pp. 221-242, 1995.

[A96]   A. Arbel, "Negative Feedback Revisited," *Analog Integrated Circuits and Signal Processing*, Vol. 10, pp. 157-178, 1996.

[AFA96] M. Abuelma'atti, A. Farooqi, S. Alshahrani, "Novel RC Oscillators Using the Current-Feedback Operational Amplifier," *IEEE Trans. on Circuits and Systems - part I*, Vol. 43, No. 2, pp. 155-157, Febr. 1996.

[AH87] P. Allen, D. Holberg, *CMOS Analog Circuit Design*, Saunders College Publishing, 1987.

[B43]   R. Blackman, "Effect of Feedback on Impedance," *BSTJ*, Vol. 22, pp. 269-277, Oct. 1943.

[B90]   D. Bowers, "Applying Current Feedback to Voltage Amplifiers," in C. Tomazou, F. Lidgey, D. Haigh (Eds.), *Analogue IC design: the current-mode approach*, IEE, 1990.

[B91]   S. Ben-Yaakov, "A Unified Approach to Teaching Feedback in Electronic Circuits Courses," *IEEE Trans. on Education*, Vol. 34, No. 4, pp. 310-316, Nov. 1991.

[B93]   D. Bowers, "The impact of New Architectures on the Ubiquitous Operational Amplifier," in J. Huijsing, R. van der Plassche, W. Sansen (Eds.), *Analog Circuit Design*, Kluwer Academic Publishers, 1993.

[B931] E. Bruun, "Feedback Analysis of Transimpedance Operational Amplifier Circuits," *IEEE Trans. on Circuits and Systems - part I*, Vol. 40, No. 4, pp. 275-277, April 1993.

[B97]   J. Bales, "A Low-Power, High-Speed, Current-Feedback Op-Amp with a Novel Class AB High Current Output Stage," *IEEE J. of Solid-State Circuits*, Vol. 32, No. 9, pp. 1470-1474, Sept. 1997.

[BAR80] W. Black Jr., D. Allstot, R. Reed, "A High Performance Low Power CMOS Channel Filter," *IEEE J. of Solid-State Circuits*, Vol. SC-15, No. 6, pp. 929-938, Dec. 1980.

[BEG74] J. Bussgang, L. Ehrman, J. Graham, "Analysis of Nonlinear Systems with Multiple Inputs," *Proc. IEEE*, Vol. 62, No. 8, pp. 1088-1119, August 1974.

[BR71] E. Bedrosian, S. Rice, "The Output Properties of Volterra Systems (Nonlinear Systems with Memory) Driven by Harmonic and Gaussian Inputs," *Proc. IEEE*, Vol. 59, No. 12, pp. 1688-1707, December 1971.

[BW83] K. Brehmer, J. Wieser, "Large Swing CMOS Power Amplifier," *IEEE J. of Solid-State Circuits*, Vol. SC-18, No. 6, pp. 624-629, Dec. 1983.

[C78]   E. Cherry, "A New Result in Negative Feedback Theory, and its Application to Audio Power Amplifiers," *Int. J. Circuit Theory*, Vol. 6, pp. 265-288, July 1978.

[C81]   E. Cherry, "Transient Intermodulation Distortion-Part I: Hard Nonlinearity," *IEEE Trans. On Acoustic Speech and Signal Processing*, Vol. ASSP-29, No. 2, pp. 137-146, April 1981.

[C821] E. Cherry, "Feedback, Sensitivity, and Stability of Audio Power Amplifier," *J. Audio Eng. Soc.*, Vol. 30, No. 5, pp. 282-305, May 1982.

[C822] E. Cherry, "Nested Differentiating Feedback Loops in Simple Audio Power Amplifier," *J. Audio Eng. Soc.*, Vol. 30, pp. 295-305, May 1982.

[C831] E. Cherry, "Amplitude and Phase of Intermodulation Distortion," *J. Audio Eng. Soc.*, Vol. 31, No. 5, pp. 298-304, May 1983.

[C832] E. Cherry, "Further thoughts on Feedback, Sensitivity, and Stability of Audio Power Amplifier," *J. Audio Eng. Soc.*, Vol. 31, No. 11, pp. 854-857, Nov. 1983.

[C90]   J. Choma Jr., "Signal Flow Analysis of Feedback Networks," *IEEE Trans. on Circuits and Systems*, Vol.37, No.4, pp. 455-463, April 1990.

[C91]   Wai-Kai Chen, *Active Network Analysis*, World Scientific, 1991.

[C93]   R. Castello, "CMOS Buffer Amplifier," in J. Huijsing, R. van der Plassche, W. Sansen (Ed.) *Analog Circuit Design*, Kluwer Academic Publishers, 1993, pp. 113-138.

[C96]   E. Cherry, "Comments on: A 100-MHz 100 dB Operational Amplifier with Multipath Nested Miller Compensation Structure," *IEEE J. of Solid-State Circuits*, Vol. 31, No. 5, pp. 753-54, May 1996.

[CC82]   E. Cherry, G. Cambrell "Output Resistance and Intermodulation Distortion of Feedback Amplifiers," *J. Audio Eng. Soc.*, Vol. 30, No. 4, pp. 178-191, April 1982.

[CD86]   E. Cherry, K. Dabke, "Transient Intermodulation Distortion-Part 2: Soft Nonlinearity," *J. Audio Eng. Soc.*, Vol. 34, No. 1/2, pp. 19-35, Jan./Febr. 1986.

[CG73]   B. Cochrun, A. Grabel, "A method for the Determination of the Transfer Function of electronic Circuits," *IEEE Trans. on Circuit Theory*, Vol. CT-20, No. 1, pp. 16-20, Jan. 1973.

[CH63]   E. Cherry, D. Hooper, "The design of Wide-Band Transistor Feedback Amplifiers," *Proc. IEE*, Vol. 110, No. 2, pp. 375-389, Feb. 1963.

[CMC94] S. Celma, P. Martinez, A. Carlosena, "Current Feedback Amplifiers Based Sinusoidal Oscillator," *IEEE Trans. on Circuits and Systems - part I*, Vol. 41, No. 12, pp. 906-908, Dec. 1994.

[CMPP93] G. Caiulo, F. Maloberti, G. Palmisano, S. Portaluri, "Video CMOS Power Buffer with Extended Linearity," *IEEE J. of Solid-State Circuits*, Vol. 28, No. 7, pp. 845-848, July 1993.

[CN91]   R. Castello, G. Nicollini, P. Monguzzi, "A High-Linearity 50-Ω CMOS Differential Driver for ISDN Applications," *IEEE J. of Solid-State Circuits*, Vol. 26, No. 12, pp. 1809-1816, Dec. 1991.

[DK69]   C. Desoer, E. Kuh, *Basic Circuit Theory*, McGraw-Hill 1969

[DM80]   A. Davis E. Moustakas, "Analysis of Active RC Network by Decomposition," *IEEE Trans. on Circuits and Systems*, Vol. CAS-27, No. 5, pp. 417-419, May 1980.

[EH92]   R. Eschauzier, J. Huijsing, "A 100-MHz 100-dB Operational Amplifier with Multipath Nested Miller Compensation," *IEEE J. of Solid-State Circuits*, Vol. 27, No. 12, pp. 1709-1716, Dec. 1992.

[EH95]   R. Eschauzier, J. Huijsing, *Frequency Compensation Techniques for Low-Power Operational Amplifiers*, Kluwer Academic Publishers, 1995.

[EHH94] R. Eschauzier, R. Hogervorst J. Huijsing, "A Programmable 1.5 V CMOS Class-AB Operational Amplifier with Hybrid Nested Miller Compensation for 120 dB Gain and 6 MHz UGF," *IEEE J. of Solid-State Circuits*, Vol. 29, No. 12, pp. 1497-1504, Dec. 1994.

[F92]   A. Fabre, "Gyrator Implementation from Commercially Available Transimpedance Operational Amplifiers," *Electronics Letters*, Vol. 28, No. 3, pp. 263-264, Jan. 1992.

[F93]   A. Fabre, "Insensitive Voltage-Mode and Current-Mode Filters from Commercially Available Transimpedance Opamps," *IEE Proc. Part G*, Vol. 140, No. 5, pp. 116-130, Oct. 1993.

[F97]   S. Franco, *Design with Operational Amplifiers and Analog Integrated Circuits*, Mc Graw-Hill, 1997.

[FF87]   G. Fortier, I. Filanovsky, "A Compensation Technique for Two-stage CMOS Operational Amplifiers," *Microelectronics Journal*, Vol.18 No.3, pp.50-57, 1987.

[FH91]   J. Fonderie, J. Huijsing, "Operational Amplifier with 1-V Rail-to-Rail Multipath-Driven Output Stage," *IEEE J. of Solid-State Circuits*, Vol. 26, No. 12, pp. 1817-1824, Dec. 1991.

[G85]   M. Ghausi, *Electronic Devices and Circuits: Discrete and Integrated*, Saunders College Publishing, 1985.

[G93]   W. Gross, "New High Speed Amplifier Design, Design Techniques and Layout Problems," in J. Huijsing, R. van der Plassche, W. Sansen (Eds.), *Analog Circuit Design*, Kluwer Academic Publishers, 1993.

[GM74]   P. Gray, R. Meyer, "Recent Advances in Monolithic Operational Amplifier Design," *IEEE Trans. on Circuits and Systems*, Vol. Circuits and Systems-21, No. 3 pp. 317-327, May 1974.

[GM93]   P. Gray, R. Meyer, *Analysis and Design of Analog Integrated Circuits (third edition)*, John Wiley & Sons, 1993.

[GPP00] G. Giustolisi - G. Palmisano - G. Palumbo - T. Segreto, "1.2 CMOS Opamp with a Dynamically-Biased Output Stage," *IEEE J. of Solid-State Circuits*, Vol. 35, No. 4, April 2000.

[GPP99] G. Giustolisi - G. Palmisano - G. Palumbo, "CMRR Frequency Response of CMOS Operational Transconductance Amplifier," *IEEE Trans. on Instr. and Meas*, Vol. 49, No. 1, pp. 137-143, Feb. 2000.

[GT86] R. Gregorian, G. Temes, *Analog MOS Integrated Circuits for signal processing*, John Wiley & Sons, 1986.

[H92] P. Hurst, "A Comparison of Two Approaches to Feedback Circuit Analysis," *IEEE Trans. on Education*, Vol. 35, No. 3, pp. 253-261, Aug. 1992.

[HL85] J. Huijsing, D. Linebarger, "Low-Voltage Operational Amplifier with Rail-toRail Input and Output Ranges," *IEEE J. of Solid-State Circuits*, Vol. SC-20, No. 6, pp. 1144-1150, Dec. 1985.

[HR80] J. Haslett, M. Rao, L. Bruton, "High-Frequency Active Filter Design Using Monolithic Nullors," *IEEE J. of Solid-State Circuits*, Vol. SC-15, No.6, pp.955-962, December 1980.

[IF94] M. Ismail, T. Fiez, *Analog VLSI Signal and Information Processing*, Mc Graw-Hill, 1994.

[JM97] D. Johns, K. Martin, *Analog Integrated Circuit Design*, John Wiley & Sons, 1997.

[K00] Wing-Hung Ki, "Signal Flow Graph Analysis of Feedback Amplifiers," *IEEE Trans. on Circuits and Systems-part I*, Vol. 47, No. 6, pp. 926-933, June 2000.

[KMG74] Y. Kamath, R. Meyer, P. Gray, "Relationship Between Frequency Response and Settling Time of Operational Amplifier," *IEEE J. of Solid-State Circuits*, Vol. SC-9, No. 6, pp. 347-352, Dec. 1974.

[KO91] H. Kuntman, S. Ozcan, "Minimisation of Total Harmonic Distortion in Active-Loaded Differential BJT Amplifiers," *Electronics Letters*, Vol. 27, No. 5, pp. 2381-2383, Dec. 1991.

[L94] M. Leach, "Fundamentals of Low-Noise Analog Circuit Design," *Proc. IEEE*, Vol. 82, No. 10, pp. 1515-1538, Oct. 1994.

[L95] M. Leach Jr., "On the Calculation of Noise in Multistage Amplifier," *IEEE Trans. on Circuits and Systems-part I*, Vol. 42, No. 3, pp. 176-178, March 1995.

[LG98] C. Laber, P. Gray, "A Positive-Feedback Transconductance Amplifier with Applications to High-Frequency High-Q CMOS Switched Capacitors Filters," *IEEE J. of Solid-State Circuits*, Vol.23, pp.1370-1378, Dec. 1998.

[LM99] K. Leung, P. Mok, W. Ki, "Right-Half-Plane Zero Removal Techniqe for Low-Voltage Low-Power Nested Miller Compensation CMOS Amplifier," *proc. ICECS'99*, 1999.

[LN86] J. Lin, J. Nevin, "A Modified Time-Domain Model for Nonlinear Analysis of an Operational Amplifier," *IEEE J. of Solid-State Circuits*, Vol. SC-21, No. 3, pp. 478-483, June 1986.

[LS94]   K. Laker, W. Sansen, *Design of Analog Integrated Circuits and Systems*, Mc Graw-Hill, 1994.

[M88]   M. Gupta (Ed.), *Noise in Circuits and System*, IEEE Press, 1988.

[MC93]   C. Motchenbacher, J. Connelly, *Low-Noise Electronic System Design*, J. Wiley Interscience, 1993.

[MG91]   J. Millman, A. Grabel, *Microelectronics (second edition)*, McGraw-Hill, 1987.

[MN89]   S. Mallya, J. Nevin, "Design Procedures for a Fully Differential Folded-Cascode CMOS Operational Amplifier," *IEEE J. of Solid-State Circuits*, Vol. 24, No. 6, pp. 1737-1740, Dec. 1989.

[MSE72] R. Meyer, M. Shensa, R. Eschenbach, "Cross Modulation and Intermodulation in Amplifiers at High Frequency," *IEEE J. of Solid-State Circuits*, Vol. SC-7, No. 1, pp. 16-23, February 1972.

[MT90]   C. Makris, C. Toumazou, "Current-Mode Active Compensation Techniques," *Electronics Letters*, Vol.26, No.21, pp.1792-1794, 11th Oct. 1990.

[MT96]   J. Mahattanakul, C. Toumazou, "A Theoretical Study of the Stability of High Frequency Current Feedback Op-Amp Integrators," *IEEE Trans. on Circuits and Systems - part I*, Vol. 43, No. 1, pp. 2-12, January 1996.

[MW95]   R. Meyer, A. Wong, "Blocking and Desensitization in RF Amplifier," *IEEE J. of Solid-State Circuits*, Vol. 30, No. 8, pp. 944-946, Aug. 1995.

[N95]   C. Neacsu, "Polynomial Analysis of Nonlinear Feedback Amplifier," *Proc. ECCTD'95*, pp. 455-458, Istanbul, 1995.

[NG93]   G. Nicollini, C. Guardiani, "A 3.3-V 800-nVrms Noise, Gain-Programmable CMOS Microphone Preamplifier Design Using Yield Modeling Technique," *IEEE J. of Solid-State Circuits*, Vol. 28, No. 8, pp. 915-920, Aug. 1993.

[NP73]   S. Narayanan, C. Poon, "An Analysis of Distortion in Bipolar Transistors Using Integral Charge Control Model and Volterra Series," *IEEE Trans. on Circuit Theory*, Vol. CT-20, No. 4, pp. 341-351, July 1973.

[NZA99] H. Ng, R. Ziazadeh, D. Allstot, "A Multistage Amplifier Technique with Embedded Frequency Compensation," *IEEE J. of Solid-State Circuits*, Vol. 34, No. 3, pp. 339-347, March 1999.

[OA90]   F. Op'T EEynde, P. Ampe, L. Verdeyen, W. Sansen, "A CMOS Large-Swing Low-Distortion Three-Stage Class AB Power Amplifier," *IEEE J. of Solid-State Circuits*, Vol. 25, No. 1, pp. 265-273, Febr. 1990.

[OS93] F. Op'Eynde, W. Sansen, *Analog Interfaces for Digital Signal Processing Systems*, Kluwer Academic Publisher, 1993.

[P93] G. Palumbo, "Optimised Design of the Wilson and Improved Wilson CMOS Current Mirror," *Electronics Letters*, Vol. 29, No. 9, pp. 818-819, April 1993.

[P96] G. Palumbo, "High-Frequency Behavior of the Wilson and the Improved Wilson MOS Current Mirror: Analysis and Design Strategies," *Microelectronics Journal*, Vol. 27, No. 1, pp. 79-85, Feb. 1996.

[P99] G. Palumbo, "Bipolar Current Feedback Amplifier: Compensation Guidelines," *Analog Integrated Circuits and Signal Processing*, Vol. 19, No. 2, pp. 107-114, May 1999.

[PC981] G. Palumbo, J. Choma Jr., "An Overview of Analog Feedback Part I: Basic Theory," *Analog Integrated Circuits and Signal Processing*, Vol. 17, No. 3, pp. 175-194, Nov. 1998.

[PC982] G. Palumbo, J. Choma Jr., "An Overview of Analog Feedback Part II: Amplifier Configurations in Generic Device Technologies," *Analog Integrated Circuits and Signal Processing*, Vol. 17, No. 3, pp. 195-219, Nov. 1998.

[PD90] M. Pardoen, M. Degrauwe, "A Rail-to-Rail Input/Output CMOS Power Amplifier," *IEEE J. of Solid-State Circuits*, Vol. 25, No. 2, pp. 501-504, April 1990.

[PM91] D. Pederson, K. Mayaram, *Analog Integrated Circuits for Communication: Principle, Simulation and Design*, Kluwer Academic Publishers, 1991.

[PNC93] S. Pernici, G. Nicollini, R. Castello, "A CMOS Low Distortion Fully Differential Power Amplifier with Double Nested Miller Compensation," *IEEE J. of Solid-State Circuits*, Vol. 28, No. 7, pp. 758-763, July 1993.

[PP01] G. Palumbo, S. Pennisi, "Current–feedback versus voltage operational amplifiers," *IEEE Trans. on Circuits and Systems-part I*, Vol. 48, No. 5, pp. 617-623, May 2001.

[PP95] G. Palmisano - G. Palumbo, "An Optimized Compensation Strategy for Two-Stage CMOS Opamp," *IEEE Trans. on Circuits and Systems -part I*, Vol. 42, No. 3, pp. 178-182, March 1995.

[PP97] G. Palmisano - G. Palumbo, "A Compensation Strategy for Two-Stage CMOS Opamps Based on Current Buffer," *IEEE Trans. on Circuits and Systems -part I*, Vol. 44, No. 3, pp. 257-262, March 1997.

[PP981] G. Palumbo - S. Pennisi, "Harmonic Distortion in Nonlinear Amplifier with Nonlinear Feedback," *International Journal of Circuit Theory and Applications*, Vol. 26, pp. 293-299, 1998.

[PP982] G. Palimsano - Palumbo, "A Novel Representation for Two-Pole Feedback Amplifiers," *IEEE Trans. on Education*, Vol. 41, No. 3, pp. 216-218, Aug. 1998.

[PP991] G. Palumbo - S. Pennisi, "Signal Amplifiers," *Encyclopedia of Electrical and Electronics Engineering*, Wiley, pp. 255-269, March 1999.

[PP992] G. Palimsano - Palumbo, "Analysis and Compensation of Two-Pole Amplifiers with a Pole-Zero Doublet," *IEEE Trans. on Circuits and Systems -part I*, Vol. 46, No. 7, pp. 864-868, July 1999.

[PP993] G. Palumbo - S. Pennisi, "Analysis of the Noise Characteristics of Current- Feedback Operational Amplifier," *Microelectronics Reliability,* Vol.40, N.2, pp. 321–327, Feb. 2000.

[PP994] G. Palumbo, S. Pennisi, "Low–Voltage Class AB CMOS Current Output Stage," *IEE Electronics Letters.* Vol. 35, N.16, pag.1329–1330, Aug. 1999.

[PPP99] G. Palmisano - G. Palumbo - S. Pennisi, *CMOS Current Amplifiers*, Kluwer Academic Publishers, March 1999.

[PPR00] G. Palmisano - G. Palumbo - R. Salerno, "CMOS Output Stages for Low Voltage Power Supply," *IEEE Trans. on Circuits and Systems -part II*, Vol. 47, No. 2, pp. 96-104, Feb. 2000.

[PPS99] G. Palmisano - G. Palumbo - R. Salerno, "1.5-V High-Drive Capability CMOS Opamp," *IEEE J. of Solid-State Circuits*, Vol. 34, No. 2, pp. 248-252, Febr. 1999.

[PT92] A. Payne, C. Toumazou, "High Frequency Self-Compensation of Current-Feedback Devices," *Proc. ISCAS'92*, pp. 1376-1379, San Diego, May 1992.

[PT96] A. Payne, C. Toumazou, "Analog Amplifiers: Classification and Generalization," *IEEE Trans. on Circuits and Systems-part I*, Vol. 43, No. 1, pp. 43-50, January 1996.

[R74] S. Rosenstark, "A Simplified Method of Feedback Amplifier Analysis," *IEEE Trans. on Education*, Vol. E-17, No. 4, pp. 192-198, Nov. 1974

[R84] S. Rosenstark, "Re-examination of Frequency response calculation for feedback amplifiers," *Int. Journal of Electronics*, Vol. 58, No. 2, pp. 271-282, 1985.

[R85]   S. Rosenstark, "Re-examination of Frequency Response Calculations for Feedback amplifiers," *Int. J. Electronics*, Vol. 58, No. 2, pp. 271-282, 1985.

[RC84]   D. Ribner, M. Copeland, "Design Techniques for Cascoded CMOS Op Amps with Improved PSRR and Common-Mode Input Range," *IEEE J. of Solid-State Circuits*, Vol. SC-19, No. 6, pp. 919-925, Dec. 1984.

[RK95]   R. Reay, G. Kovacs, "An Unconditionally Stable Two-Stage CMOS Amplifier," *IEEE J. of Solid-State Circuits*, Vol. 30, No. 5, pp. 591-594, May 1995.

[S70]   K. Simons, "The Decibel Relationship Between Amplifier Distortion Products," *Proc. IEEE*, Vol. 58, No. 7, pp. 1071-1086, July 1970.

[S91]   D. Shulman, "Speed Optimisation of Class AB CMOS OpAmp Using Doublets," *Electronics Letters*, Vol. 27, No. 20, pp. 1795-1797, Sept. 1991.

[S911]   S. Soclof, *Design and Applications of Analog Integrated Circuits*, Prentice-Hall, 1991.

[S96]   A. Soliman, "Applications of the Current Feedback Operational Amplifiers," *Analog Integrated Circuits and Signal Processing*, N.11, pp.265-302, 1996.

[S99]   W. Sansen, "Distortion in Elementary Transistor Circuits," *IEEE Trans. on Circuits and Systems-part II*, Vol. 46, No. 3, pp. 315-324, March 1999.

[SGG91] E. Sackinger, J. Goette, W. Guggenbuhl, "A General Relationship Between Amplifier Parameters and its Application to PSRR Improvement," *IEEE Trans. on Circuits and Systems*, Vol. 38, No. 10, pp. 1173-1181, Oct. 1991.

[SGH78] D. Senderowicz, P. Gray, D. Hodges, "High-performance NMOS Operational Amplifier," *IEEE J. of Solid-State Circuits*, Vol. SC-13, pp. 760-766, Dec. 1978.

[SHG78] G. Smaradoiu, D. Hodges, P. Gray, G. Landsburg, "CMOS Pulse-Code-Modulation Voice Codec," *IEEE J. of Solid-State Circuits*, Vol. SC-13, pp. 504-510, Aug. 1978.

[SKW94]   D. Smith, M. Koen, A. Witulski, "Evolution of High-Speed Operational Amplifier Architectures," *IEEE J. of Solid-State Circuits*, Vol. 29, No. 10, pp. 1166-1179, Oct. 1994.

[SM98]   H. Sjoland, S. Mattisson, "Intermodulation Noise Related to THD in Dynamic Nonlinear Wide-Band Amplifiers," *IEEE Trans. on*

Circuits and Systems-part II, Vol. 45, No. 7, pp. 873-875, July 1998.

[SPJT98] E. Seevinck, M. du Plessis, T. Joubert, A. Theron, "Active-Bootstrapping Gain-Enhancement Technique for Low-Voltage Circuits," *IEEE Trans. on Circuits and Systems - part II* Vol. 45, No.9, pp. 1250-1254, Sept. 1998.

[SS70] A. Sedra, K. Smith, "A Second-Generation Current Conveyor and Its Applications," *IEEE Trans. on Circuit Theory*, CT-17, pp.132-133, Feb. 1970.

[SS90] M. Steyaert, W. Sansen, "Power Supply Rejection Ratio in Operational Transconductance Amplifier," *IEEE J. of Solid-State Circuits*, Vol. 37, No. 9, pp. 1077-1084, Sept. 1990.

[SS91] A. Sedra, K. Smith, *Microelectronic Circuits (third edition)*, Saunders College Publishing, 1991.

[SS92] Shiyan Pei, Shu-Park Chan, "A Composite Approach for Improving the DC Performance of Current-Feedback Amplifiers," *Analog Integrated Circuits and Signal Processing*, No. 2, pp. 231-241, 1992.

[SS96] R. Senani, V. Singh, "Novel Single-Resistance-Controlled-Oscillator Configuration Using Current Feedback Amplifiers," *IEEE Trans. on Circuits and Systems - part I*, Vol. 43, No. 8, pp. 698-700, Aug. 1996.

[SW87] E. Seevinck, R. F. Wassenaar, "A Versatile CMOS Linear Transconductor/Square-law Function Circuit," *IEEE J. of Solid-State Circuits*, Vol. SC-22, pp. 366-377, June 1987.

[SY94] D. Shulman, J. Yang, "An Analitical Model for the Transient Response of CMOS class AB Operational Amplifiers," *IEEE Trans. on Circuits and Systems- part I*, Vol. 41, No. 1, pp. 49-52, Jan. 1994.

[T92] P. Tuinenga, *SPICE: A Guide to Circuit Simulation and Analysis using Pspice (second edition)*, Prentice Hall, 1992

[TG76] Y. Tsividis, P. Gray, "An Integrated NMOS Operational Amplifier with Internal Compensation," *IEEE J. of Solid-State Circuits*, Vol. SC-11, No. 6, pp. 748-754, Dec. 1976.

[TGC90] L. Tomasini, A. Gola, R. Castello, "A Fully Differential CMOS Line Driver for ISDN," *IEEE J. of Solid-State Circuits*, Vol. 25, No. 2, pp. 546-554, April 1990.

[TL94] C. Toumazou, J. Lidge, "Current Feedback OpAmps: A Blessing in Disguise?," *IEEE Circuits & Devices*, pp. 34-37, January 1994.

[TLH90] C. Toumazou, F. Lidgey, D. Haigh, *Analogue IC design: the current-mode approach*, IEE, 1990.

[TO89] F. Trofimenkoff, O. Onwuachi, "Noise Performance of Operational Amplifier Circuits," *IEEE Trans. on Education*, Vol. 32, No. 1, Feb. 1989.

[VNP80] A. Vladimirescu, A. Newton, D. Pederson, "SPICE Version 2G.1 User's Guide," Berkeley: University of California, Department of Electrical Engineering and Computer Science, 1980.

[VT92] T. Vanisri, C. Toumazou, "Wideband and High Gain Current-Feedback Opamp," *Electronics Letters*, Vol.28, No.18, pp.1705-1707, August 1992.

[W88] B. Wilson, "Constant Bandwidth Voltage Amplification Using Current Conveyor," *Int. J. Electronics*, Vol. 65, No. 5, pp. 983-988, 1988.

[W90] Z. Wang, *Current-Mode Analog Integrated Circuits and Linearization Techniques in CMOS Technology*, Hartung-Gorre Verlag Konstanz, 1990.

[W901] Z. Wang, "Analytical Determination of Output Resistance and DC Matching errors in MOS current Mirrors," *IEE Proc. Part G*, Vol. 137, No. 5, pp. 397-404, 1990.

[WG99] P. Wambacq, G. Gielen, P. Kinget, W. Sansen, "High-Frequency Distortion Analysis of Analog Integrated Circuits," *IEEE Trans. on Circuits and Systems-part II*, Vol. 46, No. 3, pp. 335-344, March 1999.

[WH95] F. Wang, R. Harjani, "An Improved Model for the Slewing Behavior of Opamps," *IEEE Trans. on Circuits and Systems-part II*, Vol. 42, No. 10, pp. 679-681, Oct. 1995.

[WH95] R. Wang, R. Harjani, "Partial Positive Feedback for Gain Enhancement of Low-Power CMOS OTAs," *Analog Integrated Circuit and Signal Processing*, No. 8, pp. 21-35, 1995.

[WHK92] J. Wait, L. Huelsman, G. Korn, *Introduction to Operational Amplifier Theory and Applications* (second ed.), Mc Graw-Hill, 1992.

[WM95] S. Willingham, K. Martin, *Integrated Video-Frequency Continuous-Time Filters High-Performance Realizations in BiCMOS*, Kluwer Academic Publishers, 1995.

[XDA00] J. Xu, Y. Dai, D. Abbott, "A Complete Operational Amplifier Noise Model: Analysis and Measurement of Correlation Coefficient," *IEEE Trans. on Circuits and Systems -part I*, Vol. 47, No. 3, pp. 420-424, March 2000.

[YA90] H. Yang, D. Allstot, "Considerations for Fast Settling Operational Amplifiers," *IEEE Trans. on Circuits and Systems*, Vol. 37, No. 3, pp. 326-334, March 1990.

[YES97] F. You, S. Embabi, E. Sanchez-Sinencio, "Multistage Amplifier Topologies with Nested Gm-C Compensation," *IEEE J. of Solid-State Circuits*, Vol. 32, No. 12, pp. 2000-2011, Dec. 1997.

ABOUT THE AUTHORS

Gaetano Palumbo was born in Catania, Italy, in 1964. He received the laurea degree in Electrical Engineering in 1988 and a Ph.D. degree from the University of Catania in 1993. In 1989 he was awarded a grant from AEI of Catania. Since 1993 he conducs courses on Electronic Devices, Electronics for Digital Systems and basic Electronics. In 1994 he joined the DEES (Dipartimento Elettrico Elettronico e Sistemistico) at the University of Catania as a researcher, subsequently becoming associate professor in 1998. Since 2000 he is a full professor in the same department.

His primary research interest has been analog circuits with particular emphasis on feedback circuits, compensation techniques, current-mode approach, and low-voltage circuits. In recent years, his research has also involved digital circuits with emphasis on high performance ones. In all these fields he is developing some research in collaboration with STMicroelectronics of Catania.

He was the co-author of the book *CMOS Current Amplifiers* published by Kluwer Academic Publishers, in 1999, and he is a contributor to the *Wiley Encyclopedia of Electrical and Electronics Engineering*. In addition, he is the author or co-author of more than 150 scientific papers on referred international journals (over 60) and in conferences. He presently is serving as an Associated Editor of the *IEEE* TRANSACTION ON CIRCUITS AND SYSTEMS part I.

Prof. Palumbo is an *IEEE Senior Member.*

Salvatore Pennisi was born in Catania, Italy, in 1965. He received the laurea degree in Electronics Engineering in 1992, and in 1997 a Ph.D. degree in Electrical Engineering, both from the University of Catania. In 1996 he joined the DEES (Dipartimento Elettrico Elettronico e Sistemistico) at the University of Catania as a researcher, where he teaches courses on Electronics and Microelectronics.

His primary research interests include CMOS analog design with emphasis on current-mode techniques. In this field he has developed various innovative building blocks and unconventional architectures for operational amplifiers. More recently, his research activities have involved low-voltage/low-power circuits, multi-stage amplifiers (with related optimised compensation techniques), and IF CMOS blocks.

Dr. Pennisi is the author or co-author of more than 60 publications on international journals and conferences, as well as of the book *CMOS Current Amplifiers* edited by Kluwer Academic Publishers, and has written an entry in the *Wiley Encyclopedia of Electrical and Electronics Engineering*.

Dr. Pennisi is an *IEEE Member.*

Printed by Books on Demand, Germany